# MICHAEL GUILLEN

# The End of Life as We Know It

## OMINOUS NEWS FROM THE FRONTIERS OF SCIENCE

# MICHAEL GUILLEN

# The End of Life as We Know It

## OMINOUS NEWS FROM THE FRONTIERS OF SCIENCE

SALEMBOOKS
an imprint of Regnery Publishing

Regnery® is a registered trademark of Salem Communications Holding Corporation

Salem Books™ is a trademark of Salem Communications Holding Corporation

Cataloging-in-Publication data on file with the Library of Congress

ISBN: 978-1-62157-672-3
Ebook ISBN: 978-1-62157-716-4

Library of Congress Cataloging-in-Publication Data

Published in the United States by
Salem Books, an imprint of
Regnery Publishing
A Division of Salem Media Group
300 New Jersey Ave NW
Washington, DC 20001
www.Regnery.com

Manufactured in the United States of America

2018 Printing

Books are available in quantity for promotional or premium use. For information on discounts and terms, please visit our website: www.Regnery.com

# ALSO BY THE AUTHOR

*Bridges to Infinity: The Human Side of Mathematics*

*Five Equations that Changed the World:*
*The Power and Poetry of Mathematics*

*Can a Smart Person Believe in God?*

*Amazing Truths: How Science and the Bible Agree*

*The Null Prophecy*

*For my beloved son, who brightens my life every day
and whose God-given gifts, I believe,
will help guide the world toward a great future.*

# CONTENTS

# ACKNOWLEDGEMENTS

I wish to thank my gifted editor-in-chief, Gary Terashita, for getting me to write this book. I love expounding on the history of science; Gary challenged me to write about the *future* of science.

I wish to thank my fellow scientists, whose stunning achievements I showcase in these pages. You toil away in laboratories all over the world, hoping your efforts will help improve the world. I earnestly pray they will.

Likewise, thank you to my fellow journalists. Getting a story right especially complex science stories—is not easy, but essential, now more than ever. I pray your professional diligence and discernment will help create a future free of hype.

Thank you also to my literary agent, Wes Yoder, who not only gets my passion for truth but helps me promulgate it through my writing. He knows as well as anyone that truth does not need defending; it needs only to be proclaimed boldly and clearly, and it will defend itself. This book, I pray, is an example of that.

Above all, I wish to thank my wife, Laurel, whose inspiring ideas and tender loving care help keep my creative fires burning brightly. Quite simply, I could not do what I do—would not be who I am—without her extraordinary input, support, and companionship.

Here's to the journey!
*Here's to the future!*

# INTRODUCTION

# GREAT OR GRIM?

*"I think that what we are up against
is a generation that is by no means sure
that it has a future."*

George Wald

This book is about the future.

It's about the unprecedented technologies and know-hows gathering momentously on the horizon. It's about the scientists and their salesmen whose hype fills us with hope. But above all, it's about you—yes, *you*—because it is *you* who will help determine if the powerful innovations described in this book will result in a great future—or a grim fate.

To anyone keeping up with the news, it's obvious science never rests. During every minute of every day, scientists somewhere in the world are wide awake, exploring, discovering, bringing to life inventions with enormous promise but also substantial risk.

Where is this vast, tireless, impressive scientific and technological prowess taking us? To a better life? To a worse life?

Can we trust scientists? Are they, as they insist, guileless seekers of truth? Or are they corrupted by politics and other self-serving agendas?

Is it wrong—anti-intellectual, anti-scientific—for society to control science? Is it even possible? Or are we fated to go along for the ride, come what may?

I am a theoretical physicist with some honest answers to offer you. I am also an award-winning science journalist who will help you digest what's going on.

For this book, I've chosen to focus on the four subject areas radically upending our lives: the world wide web, robots and artificial intelligence, surveillance technology, and genetic engineering. I devote four chapters to each subject. The first explains how the innovation came to be; the other three—relying on the very latest news from the frontiers of science—describe where it is taking us.

Before diving in, take a moment right now to see how, in just a few words, today's headlines presage the ambiguous future already within sight—a future of either monumental greatness or catastrophic grimness. Or both.

## WEB

Distance Learning Is Now Open to All Thanks to the Internet[1]
How Telemedicine Is Revolutionizing Health Care[2]

*

Facebook Says It Can't Guarantee Social Media Is Good for Democracy[3]
Your Social Media Addiction Is Giving You Depression[4]
Dating Apps Fueling Rise in Casual Sex[5]
Why a Rising Number of Criminals Are Using Facebook Live to Film Their Acts[6]
Why Are People Live-Streaming Their Suicides?[7]
Why the Internet Makes Us Monsters[8]
Former Facebook Exec Says Social Media Is Ripping Apart Society[9]

*

The Internet of Things Can Save 50,000 Lives a Year[10]

\*

Massive Ransomware Infection Hits Computers in 99 Countries[11]
The Darkening Web: Misinformation Is the Strongest Cyberweapon[12]

## ROBOT

Paralyzed Woman Moves Robotic Arm with Her Mind[13]
Bionic Eyes Can Already Restore Vision, Soon They'll Make It
Superhuman[14]
Bionic Pancreas Shows Success at Controlling Blood Sugar[15]
Google's AI Invents Sounds Humans Have Never Heard Before[16]

\*

Domino's Will Begin Using Robots to Deliver Pizzas in Europe[17]
LG Electronics to Sell Robots to Replace Hotel, Airport, Supermarket
Employees[18]
Amazon's Robot Workforce Has Increased by 50 Percent[19]
Robots Are Coming for Jobs of as Many as 800 Million Worldwide[20]
Robot Doctors Come a Step Closer as a Machine Passes Medical
Exams with Flying Colors[21]
Robot Surgeons Are Stealing Training Opportunities from Young
Doctors[22]
The Inventor of the World Wide Web Says Computers Will Someday
Run Companies without Humans[23]
Elon Musk: Robots Will Take Your Jobs, Government Will Have to
Pay Your Wage[24]
Stephen Hawking Warns Artificial Intelligence Could End Mankind[25]

\*

GE's Jeff Immelt: Robots Won't Kill Human Jobs[26]
Why Robots Will Be the Biggest Job Creators in World History[27]

\*

Humans Must Merge with Machines or Become Irrelevant in AI Age[28]
10 New Technologies that Will Make You a Cyborg[29]
Hyundai's Wearable Robots Could Make You Superhuman[30]
DARPA Is Planning to Hack the Human Brain to Let Us "Upload" Skills[31]
Godlike "Homo Deus" Could Replace Humans as Tech Evolves[32]

*

GM to Test Fleet of Self-Driving Cars In New York[33]
Self-driving Uber Car Kills Arizona Woman Crossing Street[34]
Get Ready for Freeways that Ban Human Drivers[35]
Self-Driving Cars Programmed to Decide Who Dies in a Crash[36]

*

Supersmart Robots Will Outnumber Humans within 30 Years, Says Softbank CEO[37]
Europe Mulls Treating Robots Legally as People[38]
When Machines Can Do Any Job, What Will Humans Do?[39]

## SPY

Amazon's Alexa Helps Catch Thief Red Handed[40]
Alexa, What Other Devices Are Listening to Me?[41]
London Says Media Company's Spying Rubbish Bins Stink[42]
Is Your Smartphone Listening to Everything You Say?[43]

*

Amazon May Give Developers Your Private Alexa Transcripts[44]
App Developer Access to iPhone X Face Data Spooks Some Privacy Experts[45]

*

Hundreds of Apps Using Ultrasonic Signals to Silently Track Smartphone Users[46]
How This Internet of Things Stuffed Animal Can Be Remotely Turned into a Spy Device[47]

*

No, You're Not Being Paranoid. Sites Really Are Watching
Your Every Move[48]
Facebook Can Track Your Browsing Even after You've Logged out,
Judge Says[49]
Creepy New Website Makes Its Monitoring of Your Online Behavior
Visible[50]
Mattress Startup Casper Sued for "Wiretapping" Website Visitors[51]

*

Surveillance Cameras Are Everywhere, Providing Protection—But Not
Much Privacy[52]
Creepy Website Shows Live Footage from 73,000 Private Security
Cameras Globally[53]
China's All-Seeing Surveillance State Is Reading Its Citizens' Faces[54]
Caught on Camera: You Are Filmed on CCTV 300 Times a Day in
London[55]
Americans Vastly Underestimate Being Recorded on CCTV[56]
After Boston: The Pros and Cons of Surveillance Cameras[57]
AI-Powered Body Cams Give Cops the Power to Google Everything
They See[58]
The Camera in Your TV Is Watching You[59]
Walmart Is Developing a Robot That Identifies Unhappy Shoppers[60]
Amazon Driver Caught on Video Pooping in Front of Home[61]

*

The Vast, Secretive Face Database that Could Instantly ID
You in a Crowd[62]
Smile, You're in the FBI Face-Recognition Database[63]
The New Way Police Are Surveilling You: Calculating
Your Threat Score[64]
US Navy Funds Development of Robot Surveillance System that Can
Spy on Humans in Incredible Detail[65]

*

Cyborg Dragonfly Developed for Spying[66]
Talking Drone Trying to Lure Kids from Ohio Playground[67]

Orem Police Search for Drone-Flying Peeping Tom; Takes Pics of Neighbors Through Bedroom and Bathroom Windows[68]

\*

"Mind Reading" Technology Decodes Complex Thoughts[69]
The Robot that Knows When You're Lying[70]
Algorithm Can Identify Suicidal People Using Brain Scans[71]

\*

Comey: "There Is No Such Thing as Absolute Privacy in America"[72]
DNA Scan that Can Detect 1,800 Diseases in Newborns Raises Privacy Concerns[73]
World's First "Smart Condom" Collects Intimate Data During Sex[74]
Earth's Atmosphere Can Be Turned into Massive Surveillance System Using Lasers, Scientists Discover[75]

## FRANKENSTEIN

Paralyzed People Could Walk Again Instantly after Scientists Prove Brain Implant Works in Primates[76]
Tiny Human Brains Grown in Lab Could One Day Be Used to Repair Alzheimer's Damage[77]
Scientists Implant Tiny Human Brains into Rats, Spark Ethical Debate[78]

\*

Italian Doctor Says World's First Human Head Transplant "Imminent"[79]

\*

Genetic Engineering: Way Forward for Medical Science or Sinister Threat to All Our Futures?[80]
Gene Editing Has Saved the Lives of Two Children with Leukemia[81]
A Boy with a Rare Disease Gets New Skin, Thanks to Gene-Corrected Stem Cells[82]

The Gene Editors Are Only Getting Started[83]
Top US Intelligence Official Calls Gene Editing a WMD Threat[84]

*

First Human Embryos Genetically Modified—More Will Come[85]
Chinese Researchers Announce Designer Baby Breakthrough[86]
Engineering the Perfect Baby[87]
World's First Baby Born with New "Three-Parent" Technique[88]

*

Scientists Just Took a Major Step toward Making Life from
Scratch[89]
How Scientists Are Altering DNA to Genetically Engineer New
Forms of Life[90]

*

How Upgrading Humans Will Become the Next Billion-Dollar
Industry[91]
Scientists Reverse Aging in Mammals and Predict Human Trials
within 10 Years[92]
Why Death May Not Be So Final in the Future[93]

*

Human-Pig Hybrid Created in the Lab—Here Are the Facts[94]
Organs for Human Transplant Are Being Grown inside
Sheep and Pigs[95]
A Human Ear Has Been Grown in a Rat[96]

*

Genetically Engineered Mosquitoes Could Wipe Out Zika[97]
We Might Soon Resurrect Extinct Species. Is It Worth the Cost?[98]
Peter Thiel Funding Effort to Bring Woolly Mammoths Back from
Extinction[99]
Science Fiction No More: Cloning Now an Option for
Pet Owners[100]

*

Next Phase of High-Tech Crops: Editing Their Genes[101]
What Have They Done to Our Food?[102]
Once Again, US Expert Panel Says Genetically Engineered Crops
Are Safe to Eat[103]
Would You Put the Genetically Modified Arctic Apple
in Your Pie?[104]
Scientists Convert Spinach Leaves into Human Heart Tissue—
That Beats[105]

Change isn't new. Throughout history each generation has been blindsided by one scientific game-changer or another. My wife's grandfather Bill Decatur, who turns ninety-nine this year, has watched planes, trains, and automobiles replace horse-drawn buggies; telegrams, phone calls, and emails replace letter writing; and x-rays, CAT scans, and robot surgeons replace house calls and the family doctor's trusty ol' black bag.

Nevertheless, this time is different. Very different.

Never before in human history have scientific and technological upheavals threatened to be so powerful, so intrusive, so apocalyptic. It's as if the winds of change started by previous generations have accelerated into a Category 5 hurricane. After countless centuries of great and grim changes, we are now on the brink of a future that could finally deliver the utopia we've long been pursuing—or take us out, once and for all.

## HOW I LEARNED TO AVOID HYPERBOLE

As a baby boomer, I belong to the first generation of children who grew up under the terrifying cloud of nuclear annihilation. Still, my boyhood wasn't all gloom and doom—far from it. Wide-eyed, I watched engineers build the first nuclear power plants, lasers, and computer chips; doctors discover the first oral polio vaccine and perform the first human heart transplant; and astronauts fly into space, orbit the earth, and travel all the way to the moon.

Those formative experiences taught me at least two lessons concerning the fabulous and fearsome aspects of science and technology's creations. First, it taught me to be careful not to overstate their *dangers*.

In a 1910 newspaper article reporting on a speech delivered by prominent Chicago physician Charles Gilbert Davis, the three-tiered headline screamed:

## FEARS WORLD IS GOING MAD

Dr. C. G. Davis Says Already One Man in 300 Is Insane.
Sees Doom of Civilization

What was the reason for Davis's apocalyptic concern? The evils of modern industry. "Forty thousand gaunt, hungry, exhausted children are toiling in the dust and roar of the cotton mills of the south and New England," Davis lamented. "In the great city of New York, I'm informed, 20,000 children attend school every morning suffering the pangs of hunger."[106]

The good doctor's fears, though over the top, were well founded. Yet our species managed to survive the industrial revolution and has moved on and prospered in many ways—even though, it must be said, way too many children in industrialized nations still go to school hungry and neglected.

Second, my boyhood experiences taught me to be careful not to overstate the *benefits* of science and technology's creations. In 1898 Marie and Pierre Curie discovered a mysterious, glowing element they named radium. Very quickly, fast-talking salesmen—including, alas, many scientists—hyped its miraculous powers and sold the public on a wide range of radium-laced products. From candy, toothpaste, and cosmetics to cold remedies, aphrodisiacs, and even stylish, glow-in-the-dark cocktails.[107]

"Physicians took off with the idea," reports Ross Mullner, a scientist at the University of Illinois School of Public Health in Chicago. "They tried to use it for every disease under the sun."[108]

Our very own Dr. C. G. Davis was among the physician–hypesters of the day. In a 1921 issue of the *American Journal of Clinical Medicine*, he raved: "Radioactivity prevents insanity, rouses noble emotions, retards old age, and creates a splendid youthful joyous life."[109]

It took decades before we fully realized radium's deadliness. During that time, countless people were poisoned, and many suffered gruesome deaths. Among the victims were the now-famous *radium girls*, young women who used radium paint to brush tiny numerals onto the faces of glow-in-the dark clocks and watches.[110]

In helping you, dear reader, see where science and technology are now leading us, I have striven for balance—avoiding Dr. Davis's euphoric highs and doom-laden lows.

How accurate will my analysis in this book turn out to be? My Generation-Z son will surely live long enough to find out; but even someone in his sixties is liable to see the actual outcome. I say that, because the speed at which science's stunning accomplishments are overtaking us is itself one of the great perils we face. It leaves us very little time to adequately prepare for what's coming.

There are other perils as well. Here are three I consider significant:

- **Hubris.** Even the most brilliant among my fellow scientists know far less than they let on. Every day, they toy with things they do not understand fully or in too many cases at all yet boast about improving them. As the celebrated author and social commentator E. B. White remarked, "I would feel more optimistic about a bright future for man if he spent less time proving that he can outwit Nature and more time tasting her sweetness and respecting her seniority."[111]

- **Hype.** Too many cheerleaders—notably journalists and scientists themselves—routinely exaggerate the significance of an achievement, while remaining mum about its dangers. Three reasons for such hype are:

- *Fierce competition for funding.* Most scientists doing basic research rely on the generosity of private and public patrons. These include foundations such as Rockefeller, Ford, and Sloan; government agencies such as the National Institutes of Health and National Science Foundation; and, increasingly, billionaires with specific agendas, such as Jeff Bezos, David H. Koch, and Richard Branson. As Steven Edwards at the American Association for the Advancement of Science says, "the practice of science in the 21st century is becoming shaped less by national priorities or by peer-review groups and more by the particular preferences of individuals with huge amounts of money."[112]

  The Darwinian scramble for grant monies encourages a culture of hyperbole—a lesson I learned first-hand as a grad student at Cornell. The topic for my doctoral dissertation involved kinetic theory, the science of fluids, so my thesis advisor urged me to write a grant proposal to the US Navy. He helped me word the petition to emphasize—honestly, to hype—the possibility my work could one day assist in developing new naval weaponry. It worked; I received my funding.

- *Ego.* Scientists constantly fight for the prestige and influence that comes with receiving high-profile honors, as well as having their work referenced in professional journals and the public media. This battle for scientific supremacy is every bit as cut-throat as HBO's *Game of Thrones.*

  During my thirty-plus years as a science journalist, I have witnessed the scientific ego in all its unseemly glory. Room-temperature superconductivity, for example, is an area of research that might one day (no hype) improve our ability to generate clean, electrical energy. While preparing a report on it for *Good Morning America*, my producers and I were bombarded with phone calls from rival university scientists, each faulting the other's work and lobbying to have their own research showcased on national TV.

- *Sensationalized media*. Being a physicist, I can usually keep scientists honest during media interviews. I don't let them wax hyperbolic—for example, to claim their research will one day cure all diseases and end world hunger. Regrettably, though, the average general-assignment reporter is easily snowed and thus becomes a naïve, starry-eyed enabler of scientific hype.

- **Human Nature**. Generation after generation, science offers us new, improved ways to live. But our basic, flawed nature invariably spoils or outright sabotages the opportunities. We drive cars, fly in planes, and carry smartphones, but the basics of the human experience are pretty much what they were in Old Testament times. As Felicia Day, a popular gamer and web goddess, puts it: "The internet is amazing because it connects us with one another. But it's also horrific because . . . it connects us with one another."[113]

Will everything turn out okay? Of all the questions I attempt to answer in these pages, this one is clearly the most important.

As an optimist, I cling to the hope everything will turn out okay; that science—as it so dearly wishes to do—will help lead to a model future for everyone. But honestly, it does not look like that to me right now.

That, dear reader, is why I penned this book. Think of it as a heads-up. A warning of things to come—and in some cases, as you'll see, things that are nearly or already here.

But also think of this book as a source of hope. Knowledge is power. The more you know of what's coming—of what will affect you, your children, and grandchildren—the better prepared you will be to help check the more disturbing possibilities of our scientific and technological innovations.

Is the best yet to come, as science keeps promising us? Or the worst? The answer, ultimately, is up to you and me.

# WEB

*"Oh! what a tangled web we weave . . ."*

Sir Walter Scott, Marmion

# MEMORY LANE

GREGORY: *What was that?* . . .
MAN #1: *I think it was 'Blessed are the cheesemakers.'*

Life of Brian, Scene 3, Sermon on the Mount

The web is familiar to us but clearly not well understood. For instance, many people use the terms internet and web interchangeably; yet as you will see, the two entities are quite distinct.

The web has a complex, convoluted history. Unlike the Greek goddess Athena, it didn't spring into existence fully formed. Rather, it emerged gradually and somewhat haphazardly from a hodgepodge of existing technologies, some dating back to the nineteenth century. It was as if a bunch of clever kids piecing together Legos surprised themselves one day by creating something truly amazing.

At the heart of the story is our species' instinctive desire to amplify its voice and influence far and wide—an impulse that shows itself early in life. As tiny, helpless infants firing off our first loud screech, we are startled by its sheer power and captivated by its ability to arrest people's attention.

Beginning centuries ago, that simple realization—the loudest voice in the room carries weight—drove us to invent ways of projecting our voice and influence. First by *broadcasting*, then by *computing*, and then by *networking*, innovations that in 1989 led to the conception of the world wide web.

## BROADCASTING

According to *Guinness World Records*, the intelligible range of a man's voice in perfectly still, outdoor conditions is about 590 feet.[114] In real life—where background noise makes for less-than-ideal conditions—that range is considerably reduced.

It appears the all-time distance record for unamplified speech goes to the eighteenth-century evangelist George Whitefield. According to Braxton Boren, a music technologist at New York University, Whitefield's stentorious voice was able to reach roughly 400 feet (121 meters), which equates to an audience of between 20,000 and 50,000 people. "When it is considered in the context of the hundreds of such crowds he attracted over his lifetime," Boren explains, "Whitefield probably spoke directly to more individuals than any orator in history."[115]

American painter and inventor Samuel F. B. Morse far exceeded the natural reach of Whitefield's voice by transmitting an electrical message over a long wire. In 1844, using a clever dot-dash code of his devising, he telegraphed the message "What hath God wrought" (from the Bible verse Numbers 23:23) over some forty miles, from Washington, DC to Baltimore, Maryland.[116]

Scottish-American scientist and teacher Alexander Graham Bell bested Morse by conveying the *human voice* across many miles by wire. The microphone in Bell's telephone had a diaphragm that fluttered when struck by sound waves. The fluttering membrane generated electrical ripples the way a fluttering hand in a swimming pool generates water ripples. The speaker (in effect, a reverse microphone) reconverted the electrical ripples into sound waves.

On January 25, 1915, Bell achieved history's first transcontinental phone call. It was placed from New York City to his now-famous assistant in San Francisco, some 3,400 miles away:

"Ahoy! Ahoy! Mr. Watson, are you there? Do you hear me?"
"Yes, Mr. Bell, I hear you perfectly. Do you hear me well?"
"Yes, your voice is perfectly distinct."[117]

By the 1950s, with the help of cables laid across the Atlantic Ocean, the human voice could be telephoned halfway around the world.[118] "Undersea cables, and long-distance communications in general, became the highest of high tech," observes American science-fiction writer Neal Stephenson, "with many of the same connotations as rocket science or nuclear physics or brain surgery would acquire in later decades."[119]

In the 1880s, on a completely different front, German physicist Heinrich Rudolf Hertz discovered that large electrical sparks gave off waves of electromagnetism, the way a bomb gives off shock waves.[120] The revelation immediately suggested the possibility of communicating *without* wires.

In 1901 the possibility became very real. Italian nobleman and electrical engineer Guglielmo Marconi successfully generated foot-long sparks, which produced electromagnetic waves so powerful they wafted clear across the Atlantic Ocean, from Cornwall, England, to St. John's, Newfoundland—a distance of roughly 2,100 miles.[121]

Like a carrier pigeon, the invisible waves carried a message. It consisted of just the letter S in Morse Code (dot-dot-dot), but it spoke volumes about the potential of wireless communication.[122]

Five years later, on Christmas Eve at 9:00 p.m. (EST), Canadian-American inventor Reginald Aubrey Fessenden aired the first wireless radio *voice* program. In Brant Rock, Massachusetts, Fessenden stepped up to a microphone and sent greetings to radio-ready ships on the Atlantic and Caribbean within a radius of several hundred miles. Details are a bit sketchy, but he reportedly played Handel's "Largo" on an Edison phonograph, performed "O Holy Night" on the violin, and then, after readings from the Bible, signed off with a cheery "Merry Christmas, everyone."[123]

Broadcasting technology reached a climax of sorts on April 7, 1927, when scientists publicly demonstrated a way to marry voice to moving images. Then Secretary of Commerce Herbert Hoover stood before a microphone and TV camera in Washington, DC and solemnly declared to a small audience in New York City: "Human genius has now destroyed the impediment of distance in a new respect, and in a manner hitherto unknown."[124]

Hoover was exactly right. In the two centuries since Reverend White-field's record-setting oratory, the range of our voice and influence increased from hundreds of feet to thousands of miles. An amazing achievement, to be sure. But just the beginning.

## COMPUTING

At the start of the nineteenth century, numerical tables were all the rage. Astronomers used them for navigating the night sky, ship captains for plotting courses at sea, artillery officers for positioning and aiming their massive weaponry, and tax collectors for levying tariffs. But the numerical tables—hand-calculated by minions called *human computers*—were riddled with errors.

On June 14, 1822, Englishman Charles Babbage came before the august members of the Royal Astronomical Society with a seemingly far-out solution to the problem: a hand-cranked computing device he claimed could do tedious calculations with great accuracy. Babbage's proposed contraption would require 25,000 precision-milled parts and weigh four tons.[125]

Alas, the persnickety inventor never completed the gigantic machine; and his vision of replacing humans with automated brainiacs pretty much died with him. It stayed moribund until the 1940s, when American physicist John Vincent Atanasoff at Iowa State College (now Iowa State University) and other scientists began developing rudimentary electronic computers.[126]

Their pioneering efforts inspired many separate efforts, which reached a highpoint on February 14, 1946. On that historic day, University of Pennsylvania electrical engineers publicly unveiled a thirty-ton,

1,800-square-foot programmable calculator named ENIAC, an acronym for Electronic Numerical Integrator and Computer.[127]

ENIAC's lightning-fast electronic brain—comprising one hundred thousand vacuum tubes, diodes, relays, resistors, and capacitors—could execute 50,000 instructions per second. It completed in a mere *thirty seconds* what it took the average human computer twenty hours and the best mechanical calculators of the day twelve hours to do.[128]

ENIAC and other electronic digital computers were prohibitively expensive, however, so they remained novelties well into the twentieth century. Even NASA had to rely on human computers to launch the space age. The 2017 hit movie *Hidden Figures* commemorates three such human computers—all black women—who helped calculate the flight paths for Alan Shepard's 1961 and John Glenn's 1962 history-making missions.[129]

## NETWORKING

The final leg of our winding journey toward the web was piloted by a handful of visionaries. They saw computers as much more than just fancy adding machines.

Among the prophets, those living in the United States benefitted greatly by a surprise event during the Cold War. In 1957 the Soviet Union launched the world's first satellite—a mysterious, beach-ball–sized, beeping metal sphere called Sputnik.[130]

President Dwight D. Eisenhower reacted to the threatening incident by creating the Advanced Research Projects Agency (ARPA). Psychologist and computer scientist Robert W. Taylor recalls the president was eager to boost the nation's scientific and technological prowess "so that we would not get caught with our pants down again."[131]

One of ARPA's first priorities was to improve the intolerable situation with research computers of the day. Because they were gigantic, expensive, and scarce, very few scientists had access to one.

In 1962 Joseph Carl Robnett Licklider—the first director of ARPA's Information Processing Techniques Office—floated an ingenious remedy that saw computers as elements of a telephonic grid. By using telephone lines to connect widely separated scientists and computers, he proposed, we could create an "intergalactic computer network."[132]

In a 1968 paper titled, "The Computer as a Communication Device," Licklider and Taylor further prophesied: "In a few years, men will be able to communicate more effectively through a machine than face to face. That is rather a startling thing to say, but it is our conclusion."[133]

But the vision of a telephonic computer network suffered from a glaring weakness. Legions of scientists getting on phone lines to access a small number of computers would surely create massive telephonic traffic jams.

Happily, engineers invented a device that worked like a telephonic traffic cop—the interface message processor—and *voila!* On October 29, 1969, the ARPAnet was born. It tied together computers at just four locations: UCLA, UC Santa Barbara, Stanford Research Institute (SRI), and University of Utah. But it was the true progenitor of what we now call the internet.

As a UCLA alumnus, I'm especially proud to report ARPAnet's first data transmission went from the university's engineering building, Boelter Hall, to SRI. It was an exciting moment, marked by an auspicious glitch.

UCLA computer scientist Leonard Kleinrock and his small team intended to transmit the word Login, but the fledgling net crashed prematurely and only Lo made it through. "We didn't plan it," Kleinrock recalls, "but we couldn't have come up with a better message: short and prophetic."[134]

In 1973 ARPAnet went international, by hooking up to research computers in Norway and England. Thereafter, in a kind of recapitulation of every nineteenth-century communications revolution—the telegraph, radio, and TV—computer scientists quickly found better and better ways to send *written*, *voice*, and *video* messages over the burgeoning computer network.

In 1974 scientists coined the term "internet" and created Telenet (which later became Sprintnet), the world's first commercial computer network. In 1976 Apple publicly released its first desktop computer. And on March 26 of that year Queen Elizabeth II became the first monarch to send an email. Her email address was HME2.[135]

During the following ten years, scientists further improved the internet's range and sophistication. They joined the sprawling US network to vast subnets throughout Europe and Asia. In the process, they settled on a way to assign each computer on the internet a unique number—an internet protocol (IP) address. And, on a kind of universal translator—a *transmission control protocol/internet protocol* (TCPIP)—for reconciling the growing babble of computer network languages.

Each of the incremental improvements was important. But the truly giant leap forward that produced the grand finale was taken by Tim Berners-Lee, an unassuming British computer scientist at CERN (*Conseil Européen pour la Recherche Nucléaire* or European Council for Nuclear Research), the legendary atom smasher located in Switzerland.[136] On March 12, 1989, Berners-Lee proposed a concept his supervisor reportedly belittled as "vague but interesting," in which he saw computers as *electronic libraries* and the internet as a way to access the libraries from anywhere in the world.[137]

In *Weaving the Web*, Berners-Lee recalls the struggle to find a name for his vision. Two possibilities were "information mesh" and "mine of information." He rejected the first because it sounded too much like mess and the second because its initials, moi, spelled the French word for "me," which struck him as overly possessive.[138]

He finally settled on calling it the world wide web, insisting on the three-word spelling "so that its acronym is three separate 'W's.'" Moreover, he stipulated, "There are no hyphens."[139]

In Berners-Lee's imagination, the world wide web would be a storehouse of human knowledge greater than the New Library of Alexandria, Harvard Library, and US Library of Congress combined. The www's equivalent of library books would be *websites* composed of *webpages* filled with text, sound, and video—and *hyperlinks* via which users could instantly leapfrog from one webpage to another.

The www's equivalent of catalog numbers would be addresses called URLs (uniform resource locators)—each starting with the now-familiar prefix http:// (*hypertext transfer protocol*)—that would enable users to locate precisely the information they wanted.[140]

On August 6, 1991, Berners-Lee made good on his idea, unveiling the world wide web and giving it to us free. Why gratis? "It was simply that had the technology been proprietary, and in my total control, it would probably not have taken off. The decision to make the Web an open system was necessary for it to be universal. You can't propose that something be a universal space and at the same time keep control of it."[141]

Berners-Lee's intentions were quickly realized. In 1993 *The New York Times* reported that Mosaic, the first user-friendly program designed to browse the world wide web "has grown so popular that its use is causing data traffic jams on the Internet." Mosaic would eventually give rise to today's Internet Explorer browser.

The growth of the internet was indeed explosive. In 1995 an estimated fourteen million people were online. Ten years later the number exceeded one billion. Another ten years hence more than 3.2 *billion* people worldwide were surfing the net.

In 2002 more information was stored on the web than on paper, fulfilling Berners-Lee's vision of the world wide web housing more knowledge than all the world's libraries combined. According to USC's Annenberg School for Communication and Journalism, that year "could be considered the beginning of the digital age."[142]

Today there are websites keeping constant tabs on the www's continuing growth. According to http://www.worldwidewebsize.com/, on May 1, 2018, the "Indexed World Wide Web" comprised roughly 47 billion webpages. That's how many webpages conventional search engines such as Google, Bing, and Yahoo are able to access. Upward of five hundred times more webpages exist in the so-called invisible or deep web.[143] And a particularly shadowy realm called the dark web can be plumbed only by encrypted network browsers such as TOR and I2P.[144]

What, indeed, hath God wrought?

As we'll see in the following three chapters, because of Berners-Lee's decision to give the world wide web away free, it is disrupting life today even more so than Samuel Morse's telegraph did in its day. As the respected technology journalist Cade Metz observes: "This [giveaway] allowed the web to spread, but it also allowed it to evolve in ways few could have foreseen."[145]

# STAR POWER

*"Any idiot can put up a website."*

Patricia Briggs

The world wide web differs from every other mass communication technology in two enormous ways. They are differences that fully explain why the web is at once exceptionally exciting and exceptionally frightening.

First, the web's power to amplify our voice and influence is unparalleled—far greater than that of radio or TV. To understand why, consider the basic difference between addition and multiplication.

Radio and TV enable us to reach a lot of people, each of whom is a passive receiver. So, if the reach of a broadcast swells by, say, 100 people per day, after three days the audience will be greater by 300 people: 100 + 100 + 100.

The web, likewise, enables us to reach a lot of people, but with one huge difference: each person is a receiver *and* transmitter. Each web user can share messages with others at lightning speed. So, if the reach of a broadcast grows by 100 people per day, and if each newcomer shares the message with 100 other people per day, after three days the total audience will balloon to more than *1,000,000* people: 100 × 100 × 100.[146]

The web, in other words, works like a chain letter. We mail a letter to someone, requesting he make copies and send them to, say, ten friends. Each of them, in turn, is asked to send copies to ten friends, and so forth. If everyone cooperates, the audience for the letter will multiply precipitously ($10 \times 10 \times 10 \times \ldots$). In today's lingo, the letter will go *viral*.

I routinely use the chain letter example to teach students about the power of *exponentials*—the mathematical term for anything that increases multiplicatively. Nature abounds with exponential phenomena, most notably biological cell division. In fact, the web's meaning of viral stems from the behavior of actual viruses, tiny whits of DNA with the power to take down large host organisms by multiplying uncontrollably.

The second big difference between the web and every other mass communication technology is its arrant democracy. As we've seen, Tim Berners-Lee gave away the world wide web specifically because he did not want anyone—not even himself—to control it. Ever.

The content of newspapers, magazines, books, radio, TV—all traditional forms of mass communication—is strictly controlled by an oligarchy of owners, editors, producers, and gatekeepers of various ranks who are typically well-heeled, well-educated, and well-connected. They are members of society's elite class, guardians of the establishment.

Not so with the web. In principle, every Tom, Dick, and Harry on the planet—rich, poor, educated, uneducated, it doesn't matter—is able to post content online. And at latest count, 3.7 *billion* people on all seven continents are doing exactly that, 24/7/365—for better and worse.

The result is a cacophonous, chaotic, global community the likes of which humanity has never seen before. A community of everyday people with the collective power to, among other things, transform nobodies into overnight sensations.

## LIKE A FAIRY TALE

On the radio in the 1930s and 1940s the *Original Amateur Hour*, hosted by Edward "Major" Bowes, launched the careers of Frank Sinatra, Beverly Sills, Gladys Knight, Pat Boone, Ann-Margret, and many other nascent talents. From 2002 to 2016 TV's *American Idol* did the same thing for Kelly Clarkson, Clay Aiken, Ruben Studdard, Carrie Underwood, Fantasia, Jordin Sparks, and other gifted singers.

Today the world wide web is perpetuating the tradition but at a much faster pace and on a far grander scale than ever before. Instantaneously, the web is able to confer global stardom on not just talented performers, but everyday people doing everyday things.

It began in 1991, when computer geeks at the University of Cambridge, England, pointed a video camera at their lab's communal coffeemaker and fed the live image to computers throughout their building. It gave every caffeine junky among them a fair chance at getting to a freshly brewed pot of coffee before it was completely consumed by those closest to it.

In 1993 when pioneering web browsers such as Mosaic made it easy to publish and retrieve online images, the geeks uploaded the live feed of their coffeemaker onto the world wide web and—*presto!*—XCoffee became the first viral video sensation.[147] "Only on the internet can that sort of thing happen in just a few years," remarks computer scientist Quentin Stafford-Fraser.[148]

All told, the XCoffee website was visited by hundreds of thousands of people and talked about by the world's press, including the BBC, *The Times of London*, *The Washington Post*, and *Wired*. Today, the last of the various coffeemakers made famous by the geeks—a Krups model—is on permanent display in Berlin's German Museum of Technology.[149]

In the years since XCoffee, the invention of laptops, computer pads, and smartphones has further increased the web's reach and influence by making it portable. This, in turn, has given rise to *social media*, the web-based phenomenon comprising platforms such as Facebook, Twitter, YouTube, Snapchat, and Instagram. One monumental result of social media is what I call the *triumph of the trivial*.

Consider the YouTube video of an uneducated Oklahoma woman named Kimberly "Sweet Brown" Wilkins. In April 2012, it racked up one million views in the first forty-eight hours, simply because of her colorful description of a calamitous fire in her apartment complex. Overnight, her tag line, "Ain't nobody got time for that!" was on everyone's lips.[150] The video was eventually set to music and to date has tallied more than sixty-two million views.[151]

In December 2015, Justin Bieber posted on his Instagram account the photo of a girl, together with the message: "Omg who is this!"[152] Very quickly, one of his forty-seven-plus million followers helped determine she was Cindy Kimberly, a seventeen-year-old Dutch-born student living in Spain. Quickly following her sudden stardom, Kimberly was recruited by Uno, a top modelling agency,[153] and is now a bona-fide fashion celebrity. Appearing on high-profile magazine covers and catwalks throughout the world, the once-random teenager now has an Instagram following of 3.9 million people.[154] "It seems like a fairytale," she says.[155]

Indeed.

The most popular web video of all time—"Gangnam Style"—was posted in 2012 by Psy, a little-known South Korean K-pop singer. His catchy song-and-dance routine racked up one billion views in the first five months alone and today has more than *three billion* views![156]

When Gangnam Style first came out, people of all stripes—even world leaders—were seen publicly busting out with Psy's horse-riding dance moves. Then UN Secretary General Ban Ki-Moon, who met with the singer, told him: "I hope that we can work together using your global reach . . . You have, I think, unlimited global reach."[157]

The web's unparalleled influence also makes unwilling stars out of bad actors, such as businesses that mistreat customers. "Smartphone cameras and social media have democratized information and shifted power to consumers," says Mae Anderson, tech reporter for the Associated Press (AP). "Companies can no longer sweep complaints under the rug."[158]

In April 2017, United Airlines learned this modern lesson the hard way. Passengers at Chicago's O'Hare airport recorded and posted video of an Asian American doctor, David Dao, being ignominiously bumped from their scheduled flight to Louisville to make room for a United

employee. The video—showing security officers yanking the bespectacled man from his seat and dragging him, bloodied and bruised, down
the aisle and out of the plane—went viral.

The public's outrage was instantaneous and universal. In China—
Dao's ancestral homeland and the second largest aviation market in the
world—the video quickly attracted 330 million views on Weibo and
WeChat, Chinese versions of Twitter and Facebook Messenger.[159] A
typical reaction was this post on Weibo: "The security guy beat him until
his face is covered in blood. Is this the so-called American democratic
society?"[160]

Caught off guard, United's CEO, Oscar Munoz, issued an inept
statement that was less than sympathetic to the passenger, exacerbating
the global public's fury. Munoz eventually backed down and apologized,
but it was too late; the passenger's lawyers held a press conference and
announced they were suing. A few weeks later the two parties settled out
of court.[161]

The web's stunning ability to swiftly elevate the status of everyday
people and situations is revolutionizing the retail industry as well. Historically, 1995 will always be remembered as an *annus mirabilis*—a
miraculous year—for tiny startups such as craigslist, eBay, and Amazon.

Amazon was founded by Jeff Bezos, a computer and business wonk
whose smashing success story echoes those of nineteenth-century titans
such as Andrew Carnegie, Cornelius Vanderbilt, and John D. Rockefeller.[162] The latter once remarked: "It requires a better type of mind to
seek out and to support or to create the new than to follow the worn
paths of accepted success."[163]

Bezos exemplified that "better type of mind" when he set out to
create an online bookstore. He wanted to call it Cadabra, as in abraca-
dabra, but recanted when his attorney mistook the name for cadaver. He
also considered calling it Relentless, but ultimately went with Amazon,
reportedly because he saw his business becoming as big and powerful as
the world's largest river. Indeed, his original slogan was Amazon: Earth's
Biggest Bookstore.[164] (Nevertheless, even today, enter *www.relentless.
com* and you will be directed to Amazon.)

Working out of his garage, Bezos went live with his web business in July 1995. During the first month alone, he sold books to customers in all fifty states and forty-five countries.[165] "I knew this was going to be huge," Bezos recalls. "It was obvious that we were onto something much bigger than we ever dared to hope."[166]

Despite experiencing setbacks during the turn of the millennium—the so-called dotcom bust, when a glut of ill-conceived, poorly executed online startups failed—Bezos flourished, by doggedly sticking to a winning strategy. In 2013, just after purchasing *The Washington Post*, he explained it this way: "We've had three big ideas at Amazon that we've stuck with for 18 years, and they're the reason we're successful: Put the customer first. Invent. And be patient."[167]

Clearly, Bezos's planet-sized ambitions were a perfect fit for the planet-sized web. For, as with the Sweet Brown and Gangnam Style videos, Amazon went viral, its subscriber base growing exponentially.

Today, as of this writing, Amazon is worth in excess of $700 billion.[168] That's more than Microsoft and more than twice as much as Walmart,[169] making Amazon the third most valuable company in the world, behind only Apple and Alphabet.[170] In July 2017, Bezos overtook Bill Gates to become the richest man in the world—and, not taking account of inflation, the *richest man in history*[171]—with a current net worth north of $140 billion.[172]

Happily, Amazon's success is being shared by a growing legion of mostly small businesses hawking their wares on the sprawling website—akin to the myriad shops within a mall. Today, more than two million third-party vendors sell about 50% of Amazon's total number of paid products, which comprise everything from A to Z—just as the company's beaming logo boasts.[173]

By all accounts, e-commerce generally is radically disrupting the retail landscape, the way shopping malls once did America's downtowns. As 2017 drew to a close, an article in *Fortune* led with this ominous sentence: "This year is going to go down as the worst year on record for brick-and-mortar retail."[174] In all, retailers closed more than 7,790 stores—including Radio Shack, Payless, Rite Aid, Sears, Kmart, and Gymboree.[175] Continuing the downward spiral, more than 3,800 stores are expected to shutter in 2018, including big names such as Toys R Us, Walgreens, Ann Taylor, Best Buy, and Gap.[176]

"Today, convenience is sitting at home in your underwear on your phone or iPad," says Christian Buss, an analyst for Credit Suisse Group AG. "The types of trips you'll take to the mall and the number of trips you'll take are going to be different."[177]

Traditional high-end malls continue doing well, because consumers still do like shopping in physical stores. Worldwide, fully 90 percent of all retail sales still occur in brick-and-mortar venues.[178]

Nevertheless, by some estimates, of the 1,100 malls in the United States, 400 will close in the near future. They'll need to reinvent themselves, as some are already doing—converting, for instance, into office-industrial-residential-entertainment hybrid centers—or risk being razed to the ground.[179]

On a more positive note, the web's star-making power has given a real boost to philanthropy. Once upon a time, mass mailings were the best—and for many, the only affordable—way to solicit monies for worthy causes. Not anymore.

Using websites such as GoFundMe, DonorsChoose, Booster, and Omaze, any individual or institution has the power to ask the web's 3.7 billion users for donations, at little or no cost. Can you imagine the price of trying to reach that many people with a mass mailing? Or with ads on the radio or TV?

On its Success Stories page, GoFundMe makes this claim: "Over $5 billion raised for inspiring campaigns by incredible people."[180] I encourage you to read at least a few of the stories, especially if you are in sore need of counting your blessings and having your faith in humanity restored.

One of my favorite stories—because I can readily identify with it—concerns Elijah "E-Jayy" DeVaughn, a young man reared on the mean streets of Compton, California. Overcoming many hardships, he was accepted into Harvard College on a full scholarship. E-Jayy's mom, who raised him singlehandedly, posted a request for $16,000 to help cover his ancillary expenses.[181] In four months, 160 people donated $21,633.

Another one of my favorite stories is from Kickstarter, a website dedicated to raising funds for nonprofits. In a solicitation titled "Reboot the Suit," the Smithsonian Institution requested $700,000 to restore and display the spacesuits of Alan Shepard (first American in space) and Neil Armstrong (first person on the Moon).[182] In fewer than five days, it racked up $500,000 and shortly afterward more than met its goal, raising a total of $719,779.

From Sweet Brown and Psy to Jeff Bezos and Elijah DeVaughn: these exemplars of the world wide web's star-making power perfectly illustrate the two-edged future we face, and also the fundamental reason for it. The web is and always will be inane and sublime, because—as we're about to see in stunning detail—we are both those things.

# THE WILD WILD WEB

*"We have met the enemy and he is us."*

Walt Kelly

I was driving at the speed limit on Interstate 40 in Nashville, Tennessee, when a car raced up behind me and began riding my tail. After a few uncomfortable moments, I changed lanes and the car whizzed by. Glancing over at the driver—certain it would be some testosterone-drunk teenager—I beheld a well-dressed, white-haired lady.

Someone's grandmother!

Our behavior on the highway is a lot like our behavior on the web. There is something about the two settings that brings out our dark sides.

There is a word for people behaving badly on the web: *trolls*. Like crazy drivers, trolls come in all shapes, sizes, and colors. "These are mostly normal people," explains Whitney Phillips, author of *This Is Why We Can't Have Nice Things: Mapping the Relationship Between Online Trolling and Mainstream Culture*. "You want to say this is the bad guys, but it's a problem of us."[183]

Jessica Moreno, a former Reddit executive agrees: trolls are us. "The idea of the basement dweller drinking Mountain Dew and eating Doritos isn't accurate," she says. In her experience of tracking down people who post offensive comments online, more often than not, she discovered, "They would be a doctor, a lawyer, an inspirational speaker, a kindergarten teacher."[184]

So, what explains our bad behavior on the web? There are, I believe, three major reasons.

First, there is our basic, imperfect nature, a subject of endless fascination and mystery for both science and religion. Primatologist Dian Fossey summarizes it this way: "The more you learn about the dignity of the gorilla, the more you want to avoid people."[185]

Second, there is the effect of anonymity. In real life, we don't usually blow up at people who annoy us, for fear of creating a commotion, getting slugged, being fired, or worse. "But in a car—and on the Internet— all bets are off," says sociologist Anna Akbari. "We have a vehicle for fleeing the scene, for logging off from that session. We can act without social consequence, which brings out the worst in us."[186]

Third, the world wide web is so ultrademocratic it borders on being lawless. Like a twenty-first–century Dodge City, it is populated by characters every bit as colorful (and notorious) as Cherokee Bill, Prairie Dog Dave, Fat Jack, and Cockeyed Frank. Rowdies who, as one real-life Dodge City resident recalled, "feared neither God, man, nor the devil, and [were] so reckless they would pit themselves, like Ajax, against lightning, if they ran into it."[187]

On the *wild wild web*, as in the Dodge City of yore, we see the full spectrum of human behavior on display—the good, the bad, the ugly. No one knows it better than today's teens, for whom the web is like the proverbial water cooler. It's where they hang out. "Social networking sites have created new spaces for teens to interact, and they witness a mixture of altruism and cruelty," observes Amanda Lenhart of Pew Research Center's Internet & American Life Project.[188]

The combination of our having flawed natures, hiding behind computer screens (at least, we *think* we're hiding; more about that later), and not being policed explains most of the bad and truly ugly behavior we see online.

## VANITY

In the book of "Ecclesiastes," King Solomon despairs about the human experience, crying out, "Vanity of vanities! All is vanity." Whatever you think about the Bible, the wise monarch was spot on. Working in the TV and movie industries, I've had a front row seat on the timeless spectacle of human self-importance. It's not pretty.

According to psychologists Jean Twenge and W. Keith Campbell: "Narcissism—a very positive and inflated view of the self—is on the rise." In their book, *The Narcissism Epidemic*, they cite a study of 37,000 college students, which shows "narcissistic personality traits rose just as fast as obesity from the 1980s to the present."[189]

The web is probably not entirely to blame for the trend. But as we've seen, its incomparable star-making power surely does encourage, amplify, and reward self-glorification.

In 2017 LendEDU, an online student loan business, published a revealing analysis of millennials and social media. "They use these platforms," the study observed, "to boast of their daily tidings, carefully craft their public image, and feed their egos in this interconnected digital age."[190]

On the web, egotism gets to play to a global captive audience, a temptation hard to resist. "With just a few filters, a little saturation, and a clever caption," the LendEDU study points out, "social media can make even the most average joe look like an esteemed socialite."[191]

The urge to polish our online images is so irresistible, and doing it is so easy, the web's social media experience has become as scripted and phony as a reality TV show. According to LendEDU, only 6 percent of college students' online accounts are "completely true" depictions of themselves (see WEB: DISRUPTION AND DECEPTION.) "The 15 percent that said their social media was 'not true of me at all,'" LendEDU reports, "know that they are totally fabricating their lives and have not only accepted it but are seemingly fine with it."[192]

You might say the web baits us into becoming phony politicians chasing after votes—or *likes*, to use the proper vernacular. As the Lend-

EDU study explains "If you post enough artsy, chic pictures of yourself that rack up plenty of 'likes,' then real-life accomplishments will not matter because the popularity of your social media accounts will determine your status on the social hierarchy."

Curiously, LendEDU adds, our online self-centeredness even drives us to do something seemingly illogical—vote for others. "It does not matter if Instagram users genuinely enjoy other Instagrammers' posts; the only thing that matters is that each insincere expression of emotion from you will lead to your own Instagram page gaining more status."

Nothing says online narcissism more than the *selfie*—a photo taken of and by ourselves and usually posted forthwith on the web for all to see. All too often our self-absorption lures us into photogenic but extremely dangerous poses. The result: death by selfie.

The exact numbers are hard to come by—published estimates vary wildly—but everyone agrees the number of selfie deaths is increasing. According to *Emerging Technology*, after the first eight months of 2016, the number was at seventy-three fatalities, an all-time high.[193]

On May 3, 2018, KRIV-TV aired a story about the death of sixteen-year-old Kailee Mills of Spring, Texas. While riding in a car with friends, she removed her seat belt to take a selfie. Her dad explains what happened next: "The car went off the road. She was ejected, and she died instantly.  All the other kids in the car, they had their seat belts on and they all survived with very little injury."[194]

Inexplicably, many of the fatalities happen in India.[195] For example, according to an article in *The Pioneer*, the venerable newspaper based in New Delhi, on October 3, 2007, "three young men died on the railway tracks outside Bengaluru while ostensibly attempting to take selfies with an onrushing locomotive as the background." Also, it noted "Earlier this year, in southern Bengal, five young men died trying to save one who was taking a selfie hanging off a railway door."[196]

Jesse Fox, an Ohio State University communications technology professor, summarizes the deadly and growing selfie phenomenon this way: "It's all about me. It's putting me in the frame. I'm getting attention and when I post that to social media, I'm getting the confirmation that I need from other people that I'm awesome." She adds, "so who cares if you're dangling off the side of the Eiffel Tower?"[197]

## VACUITY

Years ago, when I was at *Good Morning America*, I became good friends with a duo named The God Squad—Monsignor Tom Hartman and Rabbi Marc Gellman. One day on the show, Rabbi Gellman rightly complained that TV was helping vacuous celebrityhood upstage true heroism.

Since then, the web has made things even worse by completely gutting celebrityhood of whatever tiny bit of substance it might have had. It has done so by turning *all* of us into celebrities—at least, in our own narcissistic minds it has.

We promote ourselves endlessly by posting photos of nothing particularly remarkable: a colorful sunset, a great meal, a roadside attraction, and on and on. Fully "63% of social media," says Neil Patel, a well-known web marketing guru, "is made up of images."[198]

I call it the *triumph of the trivial* meets the *triumph of the visual*.

When we browse through the web's enormous library of images and text, research shows we invariably gravitate to the former —like kids choosing candy over broccoli every time. One survey, for instance, found "four times as many consumers would rather watch a video about a product than read about it."[199]

It's because the brain is wired to handle visual data phenomenally well. How well? The brain can process a full-sized image in just thirteen milliseconds[200]—*ten times faster* than the blink of an eye.[201]

Today's web-driven visual age reminds me of my visit years ago to Lascaux Cave in southwestern France. The unique cavern is closed to the public, but as ABC News's Science Editor, I was allowed inside for

twenty minutes (without any cameras) and will never forget the experience. Painted on the cave's earthen walls are crude renderings of humans
and animals dating back 20,000 years, long before we had a written
language. Such pictographs were a main way paleolithic humans
communicated.

I'm not suggesting the web's age of photo-centricity is hurling us all
back to the paleolithic era, but clearly it does appeal to the lizard-brained,
preliterate caveman and cavewoman in all of us. In so doing, I fear, it
discourages deep thought and reduces our complex world to a series of
online slideshows.

Pope Francis addressed this concern during the 2017 World Youth
Day. "In the social media, we see faces of young people appearing in any
number of pictures recounting more or less real events, but we don't
know how much of all this is . . . an experience that can be communicated and endowed with purpose and meaning." He admonished young
people to spend less time posting photos of their superficial, day-to-day
experiences and more time pondering the profound, eternal purpose and
meaning of their lives.[202]

It's only going to get worse, warns Cisco, the world's largest com-
puter-networking company. "Globally, IP video traffic will be 82 percent
of all consumer Internet traffic by 2020, up from 70 percent in 2015."
In other words, on the web, pictures—videos, in particular—are leaving
words in the dust.[203]

## VITUPERATION

At a conference I recently attended, Bishop T. D. Jakes recounted his
experience writing an online guest column for *The Washington Post*.
Reading it, he was aghast at the hundreds of poisonous comments posted
by readers. Many of the rants were not even on point, venturing off on
tangents having absolutely nothing to do with the column's subject.

I, too, write guest columns for major online publications—among
them, *Fox News*, *U.S. World & News Report*, and *The Christian Post*—
and have long since stopped reading the comments section. It's not that

I'm thin-skinned; I simply don't enjoy seeing people—ordinary, presumably decent people—behaving so badly.

Sadly, vituperation has never had a more accommodating, more powerful ally than the web. It's so bad we've even coined a special term for it: *cyberbullying.*

George Carlin, the late 1970s-era comedian, would be especially heartsick over this development. One of his most popular routines went like this:

> "Another plan I have is World Peace Through Formal Introductions. The idea is that everyone in the world would be required to meet everyone else in the world, formally, at least once. . . . My theory is, if you knew everyone in the world personally, you'd be less inclined to fight them in a war . . ."[204]

Carlin, I believe, was on to something. As a correspondent, I've gotten to know people in dozens of countries. To this day, whenever I read a disturbing news story about one of the nations, I empathize deeply with my friends who live there.

What Carlin had in mind is not exactly what passes for relationships on social media today. On Facebook, Instagram, or Snapchat, we do not, to quote Carlin, "look the person in the eye, shake hands, repeat their name, and try to remember one outstanding physical characteristic."[205]

Nevertheless, in this age of global connectedness—of online friends and likes—surely the web represents a step in the right direction. Surely we've increased global comity by enabling far-flung strangers to greet one another as never before, no?

Facebook CEO Mark Zuckerberg, for one, thinks so; he's been saying it from the very beginning. "People sharing more—even if just with their close friends or families—creates a more open culture and leads to a better understanding of the lives and perspectives of others," he wrote in 2012, when taking Facebook public. "We believe that this creates a greater number of stronger relationships between people and that it helps people get exposed to a greater number of diverse perspectives."[206]

Notwithstanding Zuckerberg's idealism (or crafty marketing spiel), social media's actual effect on us looks rather disastrous. In bringing us together, the web is not making the world a smaller, more understanding, more loving place. It appears to be driving us apart—like a football stadium, where rivalries are amplified, not dampened.

One such deafening amplification swamped the 2016 US Presidential campaign. Hateful posts, messages, and Tweets from all sides—Republicans, Democrats, Greens, Libertarians—amped up our differences so much that today, many months after the election, we are still at each other's throats. As a Baby Boomer, I've never witnessed anything like it.

Perhaps we will find salvation in the soaring popularity of *messaging* via apps such as WhatsApp, WeChat, Telegram, and the granddaddy of them all, Facebook's Messenger, which claims 1.2 billion active users.[207] After all, messaging, being one-on-one communication, does come closer to approximating Carlin's vision.

Pavel Durov for one, Telegram's founder, is lobbying for messaging to completely replace social media. "It's pointless and time-consuming to maintain increasingly obsolete friend lists on public networks. Reading other people's news is brain clutter. To clear out room for the new, one shouldn't fear getting rid of old baggage."[208]

I tend to agree with Durov's position. But I would not count on messaging becoming our ticket to world peace—for one simple reason.

Think about it. We live in relative peace and harmony with neighbors whose views on hot-button topics—say, abortion, homosexuality, and religion—are unknown to us. Yet, the minute we learn about them, our benign relationship is immediately placed in danger of turning into an ugly war of words.

A great deal of published research affirms this common-sense observation. In the book *Privacy Online: Perspectives on Privacy and Self-Disclosure in the Social Web*, coauthor and media psychologist Sabine Trepte explains: "With the advent of social media . . . it is inevitable that we will end up knowing more about people and also more likely that we end up disliking them because of it."[209]

As the old saying goes, familiarity breeds contempt.

## VILLAINY

Let me tell you about an incident that happened on February 28, 2017. Though not villainous, it illustrates precisely why the web is so vulnerable to malefactors.

I became aware of it early in the day, while sitting in my quiet office in Nashville, Tennessee. There was no big boom or bright flash of light; I simply noticed my favorite online news sites suddenly struggling to work properly. I kept getting error messages.

Within minutes the web was consumed by a mushroom cloud of apocalyptic-sounding complaints from around the world. Tens of thousands of websites were malfunctioning—from Apple and Airbnb to News Corp and the Securities and Exchange Commission. Also misbehaving were countless everyday devices and services that rely on the so-called Internet of Things (IoT). Everything from web-controlled light bulbs, thermostats, and home security systems to satellite radio networks, smartphone apps, and even TV remotes and computer mice.[210]

People were nonplussed, wondering: *What in the heck is going on?!*

Soon, people started directing blame at the world's largest cloud computing network, Amazon Web Services (AWS)—in particular its Simple Storage Service, or S3. Yes, Amazon doesn't just sell books and stuff; it sells time and space on sixteen, monster-sized computer facilities scattered worldwide.[211] Websites of every category use AWS: People & Society (11.55%); Arts & Entertainment (11.48%); Business & Industry (11.07%); Shopping (5.01%); and Others (60.89%).[212]

The massive outage – affecting a reported 150,000 websites – cleared up after about four hours. Two days later, Amazon came clean with an explanation:

"The Amazon Simple Storage Service (S3) team was debugging an issue causing the S3 billing system to progress more slowly than expected. At 9:37AM PST, an authorized S3 team member using an established playbook executed a command which was intended to remove a small number of servers for one of the S3 subsystems that is used by the S3 billing process. Unfortunately, one of the inputs to the command was entered incorrectly and a larger set of servers was removed than intended."[213]

Two cynical tweets pretty well sum up how everyone felt about the news. @chrisalbon remarked, "The S3 crash was caused by a single person typing in a single part of a single line of code wrong." @RedEar-Ryan said, "Oh. So a typo took down half the internet this week."[214]

The takeaway was unmistakable: the world wide web is super-sensitive to villainous attacks. There are two reasons for it.

One, the web is like an electric grid that can be crippled by a single lightning bolt—except a billion times worse, because the web is not regional, it's global.

Two, as the Amazon crash proved, just one person with a computer can disrupt our lives significantly.

It is why governments are now taking cyberattacks seriously, the way they did nuclear attacks right after World War II. Indeed, just days after his election, Donald Trump promised "to develop a comprehensive plan to protect America's vital infrastructure from cyberattacks."[215]

Hackers—cyberbullies of a particularly dangerous sort—date back to the early 1970s. Even before the web was invented, they concocted viruses with names such as Creeper, Brain, Elk Cloner, and Jerusalem to infect computers.[216]

But the web has exponentially worsened the threat. In 2000, when the web was still young, the I Love You virus—AKA the Love Bug—instantly went viral, deleting files and stealing user names and passwords from more than a million computers.[217]

Today's hackers—individuals and nations—have a dizzying number of ways to attack us via the web. In addition to viruses, there are worms, Trojan Horses, shadyware, PUPs, adware, keyloggers, RAM scrapers, botnets, backdoors, rootkits, browser hijackers, ransomware—the list goes on and on.[218]

In April 2017, a Korean undergrad released into the web a novel kind of *malware* (malicious software) that takes computer files hostage. To get their files back, victims were required to score more than 200 million points in an impossibly difficult anime video game. He eventually apologized but not before the malware infected his own computer.[219]

The proliferation of villainy on the web has, predictably, spawned an entire counter-industry offering online security software and services. Worldwide, the cybersecurity market is expected to be worth $170 billion by 2020.[220]

Still, nothing ever works perfectly—as Webroot, a prominent, Colorado-based cybersecurity firm, was recently reminded. On April 24, 2017, the company's popular antivirus software system abruptly shut down countless computers around the world. Why? Because it mistook ordinary, essential parts of Microsoft Windows for criminal activity.

Webroot's own computers became overloaded by the barrage of complaints from users, but soon enough the problem was traced to a faulty rule in the antivirus system's playbook. "The rule was removed," the company announced, "and we are in the process of rolling back all of the false positives that reside in the Webroot Threat Intelligence platform."[221]

## VIOLENCE

For years, the strident liberal talk-show host Phil Donahue lobbied unsuccessfully to air a prison execution on live TV. "I am on record as being against capital punishment," he said on the *Larry King Show* in 1999. "I am making this argument not as an ideologue, but as a person who believes that this free-press establishment has a responsibility to show this issue, which has split families."[222]

He cited as precedent a *60 Minutes* segment about euthanasia that aired a year earlier. In it, Dr. Jack Kevorkian injects fifty-two-year-old Thomas Youk—beset by Lou Gehrig's disease—with potassium chloride, the same chemical used on death-row inmates. "All movement stopped," one newspaper reporter observed, "and Youk sat dead in the chair, his mouth hanging open."[223]

"It's not necessarily murder," Kevorkian says in the segment. "It could be manslaughter, not murder. But it doesn't bother me what you call it. I know what it is. This could never be a crime in any society which deems itself enlightened."

*Enlightened.*

Today, the web is going where no TV camera has ever gone before. Pressing upon us as never before the singular questions raised by Kevorkian's and Donahue's rhetoric. Is it a sign of enlightenment to publicize films of people being killed, for any reason? Or is it evidence of our well-known weaknesses for high ratings, propagandizing, voyeurism—and worst of all, outright depravity?

I'm inclined to believe the latter, for reasons I will now explain.

The web's complicity with our darkest, most gruesome, most degenerate behavior arguably began with a group of Pakistani terrorists in 2002. After kidnapping Daniel Pearl, a reporter for *The Wall Street Journal*, they beheaded him on camera and posted the video on the web.

Since then, terrorists—most notably members of the Islamic State of Iraq and Syria (ISIS) have beheaded, torched, and drowned multitudes of innocent people, always promptly posting videos of the heinous executions on the web. Snuff films are nothing new, but the web has enabled them to go mainstream.

Things got even grislier after April 16, 2016, when Facebook launched a new app called Live—designed "to make it easier to create, share and discover live videos." In typical fashion, Zuckerberg painted a thoroughly rosy picture of the new technology, the quintessence of scientific hype:

> "Live is like having a TV camera in your pocket. Anyone with a phone now has the power to broadcast to anyone in the world. When you interact live, you feel connected in a more personal way. This is a big shift in how we communicate, and

it's going to create new opportunities for people to come together."[224]

It didn't take long before Zuckerberg's newest profit center served up live, streaming videos of a twelve-year-old girl hanging herself from a tree; an eighteen-year-old man being tortured; a thirty-three-year-old aspiring actor shooting himself in the head; a fifteen-year-old girl being gang-raped; a jilted man murdering a random, seventy-four-year-old stranger in cold blood; a father hanging himself and his eleven-month-old daughter . . . The litany is long and sickening.[225]

On Periscope—the live-video streaming app launched in 2015—a nineteen-year-old French woman livestreamed her suicide; she threw herself under a moving train.[226]

Like Zuckerberg, Periscope's founders rationalize their brainchild with high-minded rhetoric—more scientific hype:

> "Periscope was founded on the belief that live video is a pow-
> erful source of truth and connects us in an authentic way with
> the world around us. We are fascinated by the idea of discov-
> ering the world through someone else's eyes. What's it like to
> see through the eyes of a protester in Ukraine? Or watch the
> sunrise from a hot air balloon in Cappadocia?"[227]

*Right. Got it.*
*Enlightenment.*

I neither entirely blame nor exonerate the web and its cool, mega-rich hypesters. People like us—not just true monsters—are committing these heinous acts; many of them born of disappointment, despair, hopelessness, and yes, the evil streak that runs through all of us. In the words of Walt Kelly, the beloved American satirist: "We have met the enemy and he is us."

If you still do not believe it, then ask yourself who watches these despicable acts on the web? It is ordinary people, the same rubberneckers who cause horrific traffic jams on the highway because they slow down to gawk at an accident.

After that twelve-year-old girl hung herself from a tree on Facebook Live, a spokesperson for the local police department said this: "The superintendent was visibly upset when he saw the pictures of the girl and was dismayed when he learned that people were watching the incident live and no one called police."[228]

## THE WEB STRIKES BACK

If we are corrupting the world wide web with our bad behavior, then there are at least four ways the web is getting back at us.

*Perfect Memory*

Like old soldiers, as the saying goes, online content never dies. But neither does it fade away. I just checked YouTube and discovered one copy of the video of the 2014 beheading of James Foley, an American photojournalist—just one copy—has 1,160,304 views.

Even if YouTube were to delete the offensive video, it and countless other gruesome reminders of our flawed nature will always find a home inside *shock sites*, truly creepy websites specifically dedicated to jolt us. This nasty stuff used to be confined to the pages of *true crime* magazines, but now, thanks to the web, it is out there for all to see—every man, woman, and child the world over.

The web's eternal memory can also shame us personally. Just ask Mark Driscoll, the former lead pastor of the now-defunct Mars Hill Church in Seattle.

In 2014—while former pastors of the multi-campus, megachurch ministry began leveling various charges against Driscoll[229]—critics found some unseemly online posts Driscoll made back in 2000 under the pseudonym William Wallace II.[230] He had already apologized for them in 2006, but now he was forced to reiterate his contrition.

"The content of my postings to that discussion board does not reflect how I feel, or how I would conduct myself today," he said. "Over the past 14 years I have changed, and, by God's grace, hope to continue to change. I also hope people I have offended and disappointed will forgive me."[231]

But it was all for naught. On October 14, 2014, under pressure, Driscoll resigned as lead pastor and withdrew from the public eye. A few months later Mars Hill—once the third-fastest-growing mega-church in the nation—permanently closed its doors.[232]

## Fast Paced

The web's lightning speed is helping to increase the already frenetic pace of life to nearly inhuman limits.

I recall in the late 1980s, when I joined ABC-TV, a typical scene in a news segment lasted seven to ten seconds. Also, it wasn't unusual for my *Good Morning America* science reports to last five to six minutes, including some Q&A with the anchor. By the time I left ABC News, in 2002, scenes lasted a few seconds *at most* and *Good Morning America* segments lasted but a few minutes.

The web is speeding things up even more than TV ever did. In 2015 Microsoft Canada studied what effect today's instantaneous, multitasking, tweet-sized environment has on the brains and behavior of 2,000 consumers. With the help of EEGs, the researchers found "that increased media consumption and digital lifestyles reduce the ability for consumers to focus for extended periods of time." The problem is most serious among "heavy social media users," 65 percent of whom are prone to being distracted and daydreaming.[233]

How big a detrimental effect are we talking about? According to the study, the average human attention span is now eight seconds long, an historic low—down from twelve seconds in 2000. By contrast, the researchers claim, the average goldfish has an attention span of *nine* seconds.

## Mob Mentality

The web is also striking back at us by threatening the very First Amendment it was designed to protect and glorify. "An anonymous poll of the writers at *TIME*," says columnist Joel Stein, "found that 80% had avoided discussing a particular topic because they feared the online response."[234]

"I probably hold back ninety percent of the things that I want to say due to fear of being called out," a college student confessed to Conor Friedersdorf, a writer for *The Atlantic*. The student went on to describe the ugliness of today's online *call-out* culture:

> "People won't call you out because your opinion is wrong. People will call you out for literally anything. On Twitter today I came across someone making fun [of] a girl who made a video talking about how much she loved God and how she was praying for everyone. There were hundreds of comments, rude comments, below the video. It was to the point that they weren't even making fun of what she was standing for. They were picking apart everything. Her eyebrows, the way her mouth moves, her voice, the way her hair was parted. Ridiculous. I am not the kind of person to be able to brush off insults like that. Hence why I avoid any situation that could put me in that position. And that's sad."[235]

"I feel genuine fear a lot," explains Lindy West, writer for GQ.com. She is an outspoken feminist who writes about her struggle with obesity and self-image.

"Someone threw a rock through my car window the other day, and my immediate thought was it's someone from the Internet." That makes for a tragic irony, she says. "Finally we have a platform that's democratizing and we can make ourselves heard, and then you're harassed for advocating for yourself, and that shuts you down again."[236]

"If there's one fundamental truth about social media's impact on democracy it's that it amplifies human intent—both good and bad," confesses Samidh Chakrabarti, a product manager at Facebook. "At its best, it allows us to express ourselves and take action. At its worst, it allows people to spread misinformation and corrode democracy."[237]

In all candor, Chakrabarti adds, "I wish I could guarantee that the positives are destined to outweigh the negatives, but I can't."[238]

*Habit Forming*

All told, scientific studies paint a disturbing picture of our relationship with social networking sites (SNS). Especially for young people.

Consider, for instance, these key findings of a major study published in May 2017, by the Royal Society for Public Health and the Young Health Movement:[239]

- Ninety-one percent of sixteen- to twenty-four-year-olds use SNS.
- Social media has been described as more addictive than cigarettes and alcohol.
- Social media use is linked with increased rates of depression and poor sleep.
- Rates of anxiety and depression in young people have risen 70 percent in the past twenty-five years—with young people saying their favorite SNS "actually make their feelings of anxiety worse." One of them states, "I'm constantly worried about what others think of my posts and pictures."
- Overall, the most harmful SNS to young people's mental health are: Instagram, Twitter, and Facebook, in that order.

In a 2017 report published in the *International Journal of Environmental Research and Public Health*, psychologists Daria Kuss and Mark Griffiths at Nottingham Trent University confirm those findings, categorically stating that: "There is a growing scientific evidence base to suggest excessive SNS use may lead to symptoms traditionally associated with substance-related addictions."[240]

Indeed, an October 2017 survey of 5,000 public- and private-school students in England commissioned by the Headmasters' and Headmistresses' Conference (HMC) found that fully 56 percent of the eleven- to eighteen-year-olds felt "they are on the edge of addiction."[241]

Psychologist Susan Weinschenk, adjunct professor at the University of Wisconsin, explains it in terms of *dopamine*, a brain chemical that causes us to desire and to seek exciting, unpredictable experiences. "With the internet, twitter, and texting we now have almost instant gratification of our desire to seek," she says. "We get into a dopamine induced loop . . . dopamine starts us seeking, then we get rewarded for the seeking, which makes us seek more. It becomes harder and harder to stop looking at email, stop texting, stop checking our cell phones to see if we have a message or a new text."[242]

Adam Penenberg, a journalism professor at New York University, learned firsthand that other brain chemicals as well are involved in SNS addiction. They include *oxytocin*, the cuddly, feel-good hormone; and the stress-related hormones *cortisol* and *ACTH*.

As an experiment, Penenberg had his blood tested while he tweeted merrily away. "My oxytocin levels spiked 13.2%," he reports, comparable to the hormonal spike experienced by a love-struck groom at a wedding. "Meanwhile," he adds, "stress hormones cortisol and ACTH went down 10.8% and 14.9%, respectively."[243]

The addictiveness of SNS is entirely intentional, says Nir Eyal, author of *Hooked: How to Build Habit-Forming Products*. It's the outcome of a tried-and-true formula—one involving triggers, actions, rewards, and investments—used by all successful Silicon Valley titans. They do so because they realize it's not the company with the best product that necessarily wins. "Instead," Eyal says, "it's the company that holds on to the monopoly of the mind—*the habit*—that wins."[244]

Sean Parker, Facebook's first president and cofounder of Napster, agrees with Eyal. "The thought process that went into building these applications, Facebook being the first of them," he explains, "was all about: 'How do we consume as much of your time and conscious attention as possible?' And that means that we need to sort of give you a little dopamine hit every once in a while, because someone liked or commented on a photo or a post or whatever."[245] He adds candidly, "The inventors, creators—it's me, it's Mark [Zuckerberg], it's Kevin Systrom on Instagram, it's all of these people—understood this consciously. And we did it anyway."[246]

According to psychologist Jean M. Twenge, young people born between 1995 and 2012—the iGen generation, she calls them—are being deeply wounded by their addiction to social media. "Around 2012, I noticed abrupt shifts in teen behaviors and emotional states," she reports. "In all my analyses of generational data—some reaching back to the 1930s—I had never seen anything like it."[247]

Compared with previous generations, she explains, iGen-ers spend way more time at home, because "their social life is lived on their phone. They don't need to leave home to spend time with their friends."[248]

As iGen-ers mature, they're having a hard time leaving the nest, which is not at all healthy. "More comfortable in their bedrooms than in a car or at a party, today's teens are physically safer than teens have ever been. They're markedly less likely to get into a car accident and, having less of a taste for alcohol than their predecessors, are less susceptible to drinking's attendant ills," explains Twenge. "Psychologically, however, they are more vulnerable than Millennials were: Rates of teen depression and suicide have skyrocketed since 2011. It's not an exaggeration to describe iGen as being on the brink of the worst mental-health crisis in decades."[249]

As our eyes are fully opening to the utter mayhem SNS are causing us, especially young people, Silicon Valley innovators are publicly expressing contrition. It's reminiscent of the profound remorse many of my fellow physicists felt about having created nuclear weapons, especially after witnessing the death and devastation they caused in Nagasaki and Hiroshima.

Chamath Palihapitiya, a former vice president at Facebook, recently told an audience at the Stanford Graduate School of Business that he feels "tremendous guilt" about his role in creating Facebook. "The short-term, dopamine-driven feedback loops we've created are destroying how society works," he says. "No civil discourse, no cooperation; misinformation, mistruth. And it's not an American problem—this is not about Russian ads. This is a global problem."[250]

"I don't know a more urgent problem than this," says Tristan Harris, a former Google employee. He studied under behavioral psychologist Brian J. Fogg, director of the Stanford Persuasive Technology Lab, whose ideas undergird the addictive formula described by Palihapitiya. "It's changing our democracy, and it's changing our ability to have the conversations and relationships that we want with each other."[251]

"It is very common for humans to develop things with the best of intentions that have unintended, negative consequences," remarks Justin Rosenstein. He co-created Facebook's famous Like button and is now upset over what it's become.

So much so, in fact, that he has sworn off using SNS—*all apps*, actually—and welcomes a candid discussion about what he and his colleagues have intentionally and unintentionally wrought. "One reason I think it is particularly important for us to talk about this now," he says ruefully, "is that we may be the last generation that can remember life before."[252]

# DISRUPTION AND
# DECEPTION

*"If you want to liberate a country,
give them the internet."*

Wacl Ghonim

*"Google can bring you back 100,000 answers.
A librarian can bring you back the right one."*

Neil Gaiman

The web is using two kinds of muscle to strong-arm the status quo. With the first, it is successfully taking on the *establishment*. With the second, it is successfully taking on the *truth*.

*Fighting the Establishment: Political*

A flexing of the first kind of muscle was dramatically illustrated on Friday, February 11, 2011—Egypt's so-called "Friday of Wrath." After using social media to help mobilize massive street protests against Hosni

Mubarak's thirty-year-long presidency, Egyptians got what they wanted: Mubarak resigned.

"Social media have become the pamphlets of the 21st century," declares Sascha Meinrath, director of the New America Foundation's Open Technology Initiative, "a way that people who are frustrated with the status quo can organize themselves and coordinate protest, and in the case of Egypt, revolution."[253]

Closer to home, the web's muscle was instrumental in the surprising election of Donald Trump, America's first populist-outsider-billionaire president. As no one before him, Trump successfully commandeered the web's ultrademocratic showground—especially Twitter—in order to speak directly to the public, circumventing the election cycle's overtly hostile media and establishment politicians of both parties.

Harvard's Shorenstein Center on Media, Politics and Public Policy published a postmortem titled, "News Coverage of the 2016 Election: How the Press Failed the Voters." According to the study, coverage of Trump's candidacy by the establishment news media—ABC, CBS, NBC, CNN, Fox, the *Los Angeles Times*, *The New York Times*, *USA Today*, *The Wall Street Journal*, and *The Washington Post*—was fully 77 percent negative. The coverage by CBS News, the worst of the lot, was *89 percent negative.* [254]

With his adroit use of the web, Trump overcame it all—his victory in the wee hours of November 9 sending shock waves throughout the planet. The London *Daily Telegraph's* big, bold headline shouted: "Trump's American Revolution."[255]

To be clear, the web does not *create* widespread disgruntlement, like the kind that ignited the revolts in Egypt and the United States. Rather, like a planet-sized megaphone, it greatly *amplifies* the human experience—both for better and for worse. It has the unrivaled muscle to transform one person's call to arms into a collective scream, a collective scream into a wrathful movement, and a wrathful movement into a headline-grabbing coup.

"I've worked my whole life to see the power of the people come to the fore," activist Rabab Al Mahdi exclaimed after Egypt's fateful Friday of Wrath. "I never thought I would be alive to see it. It's not just about Mubarak. It's . . . about the people's power to bring about the change that no one, no one thought was possible."[256]

"We are considered flyover country, as you well know, and they don't care about us," said Scott Hiltgren, a small-business owner in Wisconsin, immediately after Trump's election. "And I think it was the silent majority that finally said, 'Enough's enough. We want a change. We don't like the way things are going.'"[257]

## Fighting the Establishment: Scientific

Like politics, science is greatly influenced by an elite ruling class, of which I am a member. I gained my place in that rarefied world by earning a BS in physics, an MS in experimental high-energy physics, and a "3-D" PhD in physics, mathematics, and astronomy.

According to the latest UNESCO Science Report, in 2013 there were 7.8 million scientists worldwide.[258] Given there are 7.1 billion people in the whole world, it means we scientists represent about 0.11% of the population.[259]

Generally speaking, most people express respect for my colleagues and me. According to the most recent Pew Research Center poll, for example, 76 percent of Americans have a "great deal" or "fair amount" of confidence in science. That's just below the military (79 percent) and way above religious leaders (52 percent) and the media (38 percent).[260]

However, there are some major wrinkles. Notwithstanding people's respect for science, they don't agree with everything it claims to be true. According to a 2015 Pew poll, the three fiercest revolts against the scientific establishment are over its oft-repeated proclamations that: 1) genetically modified (GM) foods are safe to eat, 2) foods grown with pesticides are safe to eat, and 3) climate change is mostly being caused by humans.[261]

The differences of opinion are striking. Regarding GM foods, 88 percent of scientists say they're safe; only 37 percent of the public agrees. Regarding pesticide-grown foods, 68 percent of scientists say they're safe; only 28 percent of the public agrees. Regarding anthropogenic climate change, 87 percent of the scientists agree; only 50 percent of the public agrees. (A separate, more recent Pew poll puts the public's agreement at 48 percent.)[262]

These rebellions are remarkable because they include those 76 percent of Americans who have a great deal or fair amount of confidence in science. In other words, these uprisings are not against *science* itself, but against the claim that *scientific consensuses* are unquestionable.

Understandably, the scientific establishment is alarmed by these insurrections. Like any elite ruling class, it is not accustomed to having its authority challenged and has taken to the web in an all-out campaign to "educate" the infidels.

Among those leading the charge is the National Center for Science Education, a not-for-profit advocacy group based in Oakland, California. Its elegant website trumpets this battle cry:

> "NCSE defends the integrity of science education against ideological interference. We work with teachers, parents, scientists, and concerned citizens at the local, state, and national levels to ensure that topics including evolution and climate change are taught accurately, honestly, and confidently."

But others are revolting against this orthodoxy, also by using the web. I Googled "GM foods are not safe," and instantaneously (0.70 seconds, to be precise), I was served up 1,090,000 articles, authored by people and organizations with wide-ranging credentials.

The US Senate Environment and Public Works Committee, for example, posted a report revealing that, "Over 400 prominent scientists from more than two dozen countries recently voiced significant objections to major aspects of the so-called 'consensus' on man-made global warming."[263]

Many of the dissenting scientists are among the most elite members of science's elite ruling class, including Nobel Prize winners—infidels clearly not in need of being better educated. "The distinguished scientists," explains the Senate report, "are experts in diverse fields, including:

climatology; geology; biology; glaciology; biogeography; meteorology; oceanography; economics; chemistry; mathematics; environmental sciences; engineering; physics and paleoclimatology."[264]

This web-wide war between the scientific establishment and dissenters could be likened to the Egyptian and American revolts we just discussed. But I also see similarities with the essential disagreement between Catholics and Protestants.

Catholics believe the pope to be infallible, as explained by the original Vatican Council of 1870:

> "Therefore, faithfully adhering to the tradition received from the beginning of the Christian faith . . . we teach and define as a divinely revealed dogma that when the Roman pontiff speaks EX CATHEDRA . . . he possesses, by the divine assistance promised to him in blessed Peter, that infallibility which the divine Redeemer willed his Church to enjoy in defining doctrine concerning faith or morals."[265]

To anyone tempted to challenge the pope's infallibility, the Council's decree adds these choice words of warning and condemnation: "So then, should anyone, which God forbid, have the temerity to reject this definition of ours: let him be anathema."

In the sixteenth century, Martin Luther rejected papal infallibility. To him, the Bible alone—*Sola Scriptura*—is Christendom's supreme, infallible authority. "The Word of God is greater than heaven and earth," he declared, "yea, greater than death and hell, for it forms part of the power of God, and endures everlastingly."[266]

Luther's Protestant Reformation declared that ordinary people could study the Bible for themselves and come to their own conclusions, even if they ended up disagreeing with the pope. Understandably, the Catholic church fought back tooth and nail, but Protestantism went viral and transformed the entire world—greatly aided by the web of its day, the printing press, which Johannes Gutenberg invented the century before.[267] "Once the Bible was printed, people who could read had direct access to the text itself," explain the curators of the website *Musée virtuel du Protestantisme*. "This was truly a revolution."[268]

Today, those who use the web to defend the infallibility of scientific consensuses are behaving like Catholics. To them, anyone who dares—who has the "temerity"—to question the establishment's interpretations of the available evidence on such matters as GM foods, pesticides, and climate change are anathema.

Yet, as I see it, the dissenters are simply behaving like Protestants. They respect science—or should, in my opinion—but strongly disagree with the dogma that, like the pope, it is infallible. As support, they cite the many times in history when scientific orthodoxies have been upended by mutinous ideas—documented brilliantly by philosopher Thomas Kuhn in *The Structure of Scientific Revolutions* but also on the world wide web, for all to read.[269]

Albert Einstein, for one, was a scientific Protestant. His provocative ideas about space and time flew in the face of the scientific consensus of his day and were, therefore, considered anathema.

Long after his spectacular vindication, he explained why science should never be treated as infallible, no matter how strongly it believes the facts support its preferred position. "No amount of experimentation can ever prove me right," he observed, "a single experiment can prove me wrong."[270]

As a scientific Protestant myself, I agree with Einstein. Treating science as infallible is tantamount to making it into something it is not meant to be, and most assuredly does not wish to be: a dogmatic religion. I'll have more to say on the subject in my concluding chapter, SCIENCE, SACREDNESS, AND SELF-DESTRUCTION.

For now, I like how New York University physicist Steven Koonin, who served in the Obama administration's Department of Energy, explains it:

> "Consensus statements necessarily conceal judgment calls and debates and so feed the 'settled,' 'hoax' and 'don't know' memes that plague the political dialogue around climate

change. We scientists must better portray not only our cer-
tainties but also our uncertainties, and even things we may
never know. Not doing so is an advisory malpractice that
usurps society's right to make choices fully informed by risk,
economics and values. Moving from oracular consensus statements to an open adversarial process would shine much-
needed light on the scientific debates."[271]

*Fighting the Truth: Accidental*

The cancer of misinformation is nothing new: rumors and propaganda are a staple of the human experience. But today, thanks to the
web, the deceptive noise level is at an all-time high, earning it a derisive
nickname: *fake news.*

The proliferation of misinformation is a top concern of the web's
creator, Tim Berners-Lee. As he sees it, social media sites and search
engines are among the biggest malefactors:

> "These sites make more money when we click on the links
> they show us. . . . The net result is that these sites show us
> content they think we'll click on—meaning that misinforma-
> tion, or 'fake news,' which is surprising, shocking, or designed
> to appeal to our biases, can spread like wildfire."[272]

I think Berners-Lee is right to finger the distributors of fake news.
But surely the *creators* of fake news are equally to blame, even when it's
done accidentally.

Case in point: on April 14, 2017, at 10:27 pm EDT, Bloomberg's
website flashed the arresting headline: "North Korea Fires Projectile,
Media Says: Xinhua." Underneath were the words: "Story developing
. . ." In less than *sixty seconds* Blomberg amended the headline—still
datelined 10:27 PM EDT—to read: "North Korean Missile Seen at
Military Parade: Xinhua."[273]

The financial giant's honest mistake, which instantly went viral,
caused first a wave of panic and then a wave of outrage. Afterward,

Mediaite published a critical article titled "Bogus Headlines on Bloomberg and Chinese Media Risk Starting WWIII."[274]

*Fighting the Truth: Intentional*

Far more problematic is *intentional* fake news, the perpetrators of which have motives running the spectrum from noble to nefarious. Let me give you three quick examples.

- On June 16, 2006, when YouTube was but sixteen months old, a brainy, attractive sixteen-year-old homeschooled girl named Bree began a video diary called lonelygirl15. It went viral and very quickly shot to the top of YouTube's most viewed list.

  The web community's love affair with Bree was so intense, her doings and sayings were covered by the establishment media. "She has huge online fame," wrote *The New York Times* columnist Virginia Heffernan. "What could be more 2006?"[275]

  The enigmatic girl's true confessions—delivered to a webcam in her bedroom—mixed routine teenage angst (one post was titled "My Parents Suck") with deep musings about science, religion, history, and philosophy. "Today for our installment of 'Proving Science Wrong,'" she announced at the start of her sixteenth posting, "our topic is going to be the Tolstoy Principle. Jared Diamond came up with this one."[276]

  Imagine the world's reaction when Richard Rushfield, then entertainment editor for latimes.com, exposed lonelygirl15 as a hoax. "The team behind the lonelygirl15 YouTube mystery has come forward, claiming that lonelygirl15 is part of their 'show' and thanking their fans effusively for tuning in to 'the birth of a new art form.'"[277]

  *Ah, yes, deception—a new art form.*

lonelygirl15's co-creators—an eclectic group of wan-nabe screenwriters, filmmakers, actors, and a software engineer—seem to have missed the irony of calling decep-tion a new art form. In fact, they were quite defensive about what they'd done. "We never lied," says co-creator Miles Beckett, "we just put it out there. When people asked us if it was real or not, we never responded, we just let it ride."[278]

The popularity of lonelygirl15 helped seal YouTube's already rising star. A mere one month after the media's unmasking of lonelygirl15 as fake news, Google pur-chased YouTube for $1.65 billion.[279]

- On April 13, 2017, *The Huffington Post* published an article titled "Could It Be Time to Deny White Men the Franchise?" authored by a South African feminist phi-losophy student named Shelley Garland. "If white men no longer had the vote, the progressive cause would be strengthened," Garland wrote. "At the same time, a denial of the franchise to white men, could see a redistribution of global assets to their rightful owners."[280]

Predictably, the article went viral.

The following day, Verashni Pillay, editor-in-chief of *The Huffington Post's* South African bureau, defended the article in a commentary piece titled "This blog On White Men Is Going Viral. Here's Our Response." In it, she declared: "Those who have held undue power granted to them by patriarchy must lose it for us to be truly equal. This seems blindingly obvious to us."[281]

But it was a different story when, four days later, Shel-ley Garland was outed as Marius Roodt, a researcher at the Centre for Development and Enterprise, a leading South African think tank. The Huffington Post retracted the story, and Pillay sang a different tune. "I did not make it clear enough in my initial response that I absolutely do not agree with the disenfranchisement of any group of people," she said. "I don't hate white men."[282]

Despite the initial blunder and subsequent mea culpa, neither Pillay nor anyone else at The Huffington Post was fired or, it appears, even disciplined. Nevertheless, the reputation of the well-known, extremely liberal publication took a big hit, as the web lit up with comments such as these:[283]

Congrats Huffingtonpost! You are now Officially the most racist website online!

Let's do an experiment: 1 - copy this article in MS word. 2 - Hit "Ctrl+R" and replace "White men" with any other group, gender or ethnic background. 3 - Read the article again . . . Are you offended? If you are, why are you printing this garbage?

Here's a correction. Delete your entire website. You absolute psychopaths.

- Finally, consider 20th Century Fox's marketing campaign for its 2017 horror movie "A Cure for Wellness." The studio published seemingly real online, big-city newspapers— *Houston Leader, NY Morning Post, Indianapolis Gazette,* etc.—cleverly intermixing movie ads with sensational, left-leaning, fake news stories such as these:[284]

"BOMBSHELL: Trump and Putin Spotted at Swiss Resort Prior to Election"

"LEAKED: Lady Gaga Halftime Performance to Feature Muslim Tribute"

"California Legislature to Consider Tax Rebates for Women Who Get Abortions"

Reporters at BuzzFeed outed the deception just before the movie's opening weekend. Their article, "A Hollywood Film Is Using Fake News to Get Publicity," revealed that Fox's "fake stories—such as one about Donald

Trump implementing a temporary ban on vaccinations—have been picked up by real websites and generated significant engagement on Facebook thanks to people being fooled."[285]

Fox eventually apologized, but there were no reports of anyone being fired or disciplined. Moreover—no surprise, given the web's infamously long memory—the sham newspaper links remain permanently archived on the web. See for yourself by going to, say, http://archive.is/OxyG1.[286]

Another of the wellness movie's phony promotional websites, healthcuregov.com—made to look like the Affordable Care Act's official website, healthcare.gov—still lives as well but now sends you to a website peddling the studio's upcoming movies.

"This absolutely crosses the line," says Bonnie Patten, executive director of truthinadvertising.org. "Using a fake news site to lure consumers into buying movie tickets is basically a form of deceptive marketing."

It is that, of course, but much more. "I think this is a hot enough subject that most marketers would understand that taking advantage of a vulnerable public is dangerous," observes Susan Credle of FCB, a global ad agency. "When you start to tear down media and question what's real and what's not real, our democracy is threatened."[287]

Inevitably, some try putting a positive spin on the hoax—for example, by comparing it to the 1938 War of the Worlds radio broadcast, in which twenty-three-year-old Orson Welles played a reporter breaking the story about a Martian invasion. But unlike Fox's marketing campaign, the broadcast, while realistic, was up front about what it was doing.

As explained by Snopes, the well-known watchdog website, "the ['War of the Worlds'] show was a regularly scheduled and announced episode of *The Mercury Theatre on the Air*, a radio program dedicated to presenting dramatizations of literary works."[288] *War of the Worlds*, of course, is the classic science fiction novel by H. G. Wells, first published in 1897.[289]

Some find comfort in the possibility that online hoaxes are making us more cautious, less dupable—clearly, a good thing. "I think it would be really hard to do something that feels real but isn't real and have people be fooled anymore," muses Miles Beckett, co-creator of lonelygirl15. "There's been tons of hoaxes and other things since lonelygirl, and I think they've been figured out pretty damn fast because people are skeptical."

But even if it's true, our increased skepticism is now coupled with increased confusion. That was vividly illustrated in April 2017, when news stories reported Syria had dropped chemical weapons on the rebel-held town of Khan Sheikhoun. "Dozens of people, including children, died," reported *The New York Times*, "after breathing in poison that possibly contained a nerve agent or other banned chemicals, according to witnesses, doctors and rescue workers."[290]

Russia called it fake news, claiming the deadly chemicals were actually released from in situ *rebel storehouses* bombed by the Syrians. In a story posted on the web by *Pravda*, a former Department of Defense official declared: "As the United States and Western allies march closer to full-scale conflict with Syria, many of their claims are now being scrutinized and dissected by a skeptical public."

Several days later, the United States retaliated by taking out the airstrip from which the chemical attacks were allegedly launched. However, respected mainstream news sites, appearing to buy into Russia's version of events, published stories defending Syria.

At one surreal point, like a serpent feeding on its own tail, a Bloomberg story titled "Russia Says Evidence Growing Syria Chemical Attack Was Staged" actually quoted Russian Foreign Minister Sergei Lavrov referencing the barrage of pro-Syria news articles as evidence supporting Russia's position. "Publications including in the U.S. and the U.K." he said, "have highlighted 'many inconsistencies' in the version of events in Syria's Idlib province that was used to justify the American airstrikes."[291]

Similar claims made by Syrian president, Bashar al-Assad, were also run by mainstream news sites, including CNN.[292] Confusing matters further, on the day of the chemical attacks, Syria's First Lady, Asma Assad, took to Instagram, posting a photo of herself smiling benignly.

Predictably, many users roasted her for her seeming indifference—one of them accusing her husband of being "a baby killer." But others, thoroughly caught up in the confusion, questioned the very authenticity of the posting. "How stupid are the people who think the First Lady posted this picture of herself," one user said, "or that this is actually her account."[293]

The debate didn't end there. Users were left to wonder about the motives of the people posting criticisms or defenses of the First Lady, not to mention the larger controversies about the chemical attack and America's retaliation. The endless speculation was enough to make anyone's head spin. Who was lying and who was telling the truth?!

The web makes it too easy for anyone—*anyone*—to proliferate uncertain information and outright fake news, to warp reality for their own purposes, noble or nefarious. As a result, we are now like people trying to navigate through a house of mirrors, where you can't trust even *yourself* to know what to believe. Where reflections lure you into dead ends and often appear to extend to infinity, like the endless assertions, counter-assertions, and ensuing arguments commonly seen on the web.

"If people want to have stunts, fine," says Richard Edelman, CEO of his well-respected, eponymous global marketing firm, "but one of the great dangers it seems to me at the moment is people can't differentiate between that which is real and that which is a fake story."

## CLOSING THOUGHTS

Today, internet curators are trying many different ways to scrub fake news from the web—to bring law and order to Dodge City. Mostly, they fall into three broad categories.

*Establishment Solutions*

What I call establishment solutions rely on some form of top-down regulation. For example, Google's Fact Check initiative, launched worldwide in April 2017. Here's how the tech giant describes it:

> ". . . we're making the Fact Check label in Google News available everywhere and expanding it into Search globally in all languages. For the first time, when you conduct a search on Google that returns an authoritative result containing fact checks for one or more public claims, you will see that information clearly on the search results page."[294]

Some 115 hand-selected third-party reviewers—a very elite group—are given the extreme privilege of passing judgment on nearly every news story and website on the world wide web. The problem with this top-down approach, of course, is that the self-appointed fact checkers themselves invariably have biases.

On January 9, 2018, Mediaite, a left-leaning, online media watchdog magazine, published a critique of Google's Fact Check titled, "Why Does Google's New 'Fact-Check' Feature Seem to Be Targeting Conservative Sites?"[295] The author reports on the results of an impartial experiment he performed on Fact Check, with both conservative and liberal news sites.

Of the ten prominent conservative sites—*The Daily Caller, The Daily Wire, The Federalist, Breitbart, The Washington Free Beacon, TheBlaze, Townhall.com, RedState, The Weekly Standard,* and *National Review*—he found "five of them were scrutinized with the 'Reviews Claimed' tab."[296] The reviews claimed the sites were unreliable for one reason or another.

Of the ten prominent liberal sites—*Slate, Salon, Daily Kos, Media Matters for America, MoveOn.org, ThinkProgress, Mother Jones, Vice, Vox,* and *The Huffington Post*—none was flagged with any 'Reviews Claimed' tabs. *Not a one.*

"Have conservative sites been guilty of spreading misinformation? Sure. But according to Google, liberal sites haven't," concludes author Joseph A. Wulfsohn. "Perhaps the search engine will blame some nonhuman 'algorithm' for the seemingly blatant bias if they're pressed on it, but if they're going to fight 'fake news,' which frankly is a bipartisan problem, they need to have bipartisan treatment of these sites."[297]

On April 2, 2018—in celebration of the second annual International Fact-Checking Day[298]—the Monmouth University Polling Institute reported more than three of four Americans believe the mainstream media "engage in reporting fake news." Patrick Murray, the institute's director, remarked: "These findings are troubling, no matter how you define 'fake news.' Confidence in an independent fourth estate is a cornerstone of a healthy democracy. Ours appears to be headed for the intensive care unit."[299] The matter has become so grave, in 2018 Congress held investigative hearings about fake news, liberal political bias, and faulty algorithms—the one featuring Mark Zuckerberg creating quite a commotion.[300] Moreover, the Media Research Center, a conservative group, launched the "Fact-Checking the Fact-Checkers" project.[301]

Not to be outdone, California is threatening to go even further by enacting a particularly strict—some might say, Draconian—version of the establishment approach. On February 17, 2017, state assemblyman Ed Chau introduced the California Political Cyberfraud Abatement Act, which would *criminalize* people who purvey fake news:

> "The bill would also make it unlawful for a person to knowingly and willingly make, publish or circulate on a Web site, or cause to be made, published, or circulated in any writing

posted on a Web site, a false or deceptive statement designed to influence the vote on any issue submitted to voters at an election or on any candidate for election to public office."[302]

On April 19 the assembly struck the above clause from the bill, effectively gutting it.[303] But I doubt we've heard the Golden State's last word on the matter.

*Populist Solutions*

What I call the populist solutions rely on *we the people* to police the web. It is an approach very much in line with the web's ultrademocratic, freedom-loving nature—and one that seems to work. Consider two quick examples.

- On April 18, 2017, the AP tweeted this breaking story: "Fresno police say suspect in triple slaying told them he hates white people, shouted 'God is great' before killings." Immediately, people took to Twitter and pointed out the shooter actually shouted *"Allahu Akbar,"* the familiar battle cry of radical Islamic terrorists, meaning *"Allah* is the greatest."[304]

  Caught with its pants down, the AP tweeted a correction on the following day:

  > "We deleted a tweet about a Fresno slaying suspect shouting 'God is great.' It failed to note he said it in Arabic. A new tweet is upcoming."[305]

  It was a clear-cut case of the public successfully recognizing and correcting fake news. Still, it's puzzling to me why—whether out of ignorance or political correctness—the AP still failed to report the actual words spoken by the suspect, *"Allahu Akbar."*

- On April 19, 2017, *The New York Times* sports editor, Jason Stallman, posted this tweet, along with two photos that seemed to show Obama had attracted a bigger crowd than Trump:

  > "Patriots' turnout for President Obama in 2015 vs. Patriots' turnout for President Trump today."[306]

  Just four hours later, the New England Patriots tweeted information that immediately discredited Stallman's claim:

  > "These photos lack context. Facts: In 2015, over 40 football staff were on the stairs. In 2017, they were seated on the South Lawn."[307]

  It isn't clear if Stallman's deception was innocent or intentional, but he quickly apologized. "Bad tweet by me. Terrible tweet," he confessed. "I wish I could say it's complicated, but no, this one is pretty straightforward: I'm an idiot."[308]

  Once again, the public scored a victory against fake news.

## Hybrid Solutions

For the past few years, Facebook has experimented with strategies that are part establishment and part populist. "In the public debate over false news, many believe Facebook should use its own judgment to filter out misinformation," says Samidh Chakrabarti, Facebook's product manager for politics and elections. "We've chosen not to do that because we don't want to be the arbiters of truth, nor do we imagine this is a role the world would want for us."[309]

Starting in early 2017, Facebook users were allowed to flag news stories as fake or suspicious—and so were Facebook's robots, mathematical algorithms designed to recognize stinkers. If enough users and bots flagged a story, it was sent to human fact-checkers for final evaluation—entities such as Snopes, Politifact, FactCheck.org, ABC News, AP, and even *The Weekly Standard*, a conservative magazine. If enough fact-checkers gave the flagged story a thumbs down, Facebook adorned it with a banner proclaiming it to be "Disputed."[310]

On December 20, 2017, however, Facebook announced it was scrapping the whole Disputed banner thing, explaining: "Academic research on correcting misinformation has shown that putting a strong image, like a red flag, next to an article may actually entrench deeply held beliefs—the opposite effect to what we intended."[311]

Ah, yes, unintended consequences strike again.

In place of the banners, Facebook is now sending alerts to users who share disputed stories, in hopes of discouraging their dissemination. Also, they are attaching "Related Articles" to disputed stories—web documents that offer alternative points of view. "Indeed," reports Facebook, "we've found that when we show Related Articles next to a false news story, it leads to fewer shares than when the Disputed Flag is shown."[312]

Zuckerberg now admits publicly that Facebook and SNS generally have created and amplified a dystopia that must be fixed. "The world feels anxious and divided, and Facebook has a lot of work to do," he confesses. "My personal challenge for 2018 is to focus on fixing these important issues. . . . If we're successful this year, then we'll end 2018 on a much better trajectory."[313]

In my opinion, Zuckerberg is on the right track; hybrid solutions are the best way to go. I say that because once upon a time earnest, just-minded authority figures—Wyatt Earp and Bat Masterson—joined forces with an alert public to successfully civilize Dodge City.[314] Considering human nature hasn't changed a bit since the days of the wild wild west, I'm tempted to suppose it can also work on the wild wild web.

Tim Berners-Lee appears to agree. "It has taken all of us to build the web we have," he said on the web's twenty-eighth anniversary, "and now it is up to all of us to build the web we want—for everyone."[315]

My only hesitation comes from realizing that in its wildest days, Dodge City had fewer than one thousand residents. The web, on the other hand, has 3.7 *billion* inhabitants and counting[316]—each of whom, remember, is a receiver and a transmitter. No one—no ruler in history—has ever attempted to manage the behavior of that many people.

# ROBOT

*"I visualize a time when we will be to robots what dogs are to humans . . ."*

Claude Shannon

# MEMORY LANE

*"We're fascinated with robots because they are reflections of ourselves."*

Ken Goldberg

We are in the early years of what's being called the fourth industrial revolution. The first revolution was powered by steam, the second by electricity, and the third by computers. Now, the fourth revolution is being powered by science-fiction–like efforts to create artificial life and artificial intelligence—its apparent ultimate goal being the perfect integration of man and machine.

"In this fourth revolution, we are facing a range of new technologies that combine the physical, digital and biological worlds," says *Forbes* contributor Bernard Marr. "These new technologies will impact all disciplines, economies and industries, and even challenge our ideas about what it means to be human."[317]

"The fourth industrial revolution can be described as the advent of 'cyber-physical systems' involving entirely new capabilities for people and machines," says Nicholas Davis, head of Society and Innovation at the World Economic Forum.[318] "The changes are so profound," adds

Klaus Schwab, the Forum's founder, "that, from the perspective of human history, there has never been a time of greater promise or potential peril."[319]

One of the fourth revolution's greatest promises is artificial intelligence that will help cure diseases, liberate us from tedious and dangerous chores, and even predict the future. Its greatest perils are the unintended consequences destined to arise and produce calamities we can't presently imagine—as we've seen happen time and again with other novel, greatly hyped technologies.

With such profound changes in the offing, we are being forced to reconsider who or what we are, exactly. Are we nothing more than complicated machines? Can we be replicated and therefore replaced? Worse, are we in danger of becoming *subjugated* by robots and other superlative machines? Are we already slaves to our own inventiveness?

Ironically, such human-centered introspections are precisely what produced the fourth revolution in the first place. "At bottom, robotics is about us," observes Roderic Grupen, a computer scientist at the University of Massachusetts, Amherst. "It is the discipline of emulating our lives, of wondering how we work."[320]

In other words, the rise of the fourth revolution is at its core a story about self-fascination, imitation, and innovation. A story that began centuries ago, when we set about to create likenesses of ourselves that were at first *subhuman*, then *superhuman*, and ultimately *transhuman*.

## SUBHUMAN

The seventeenth-century Frenchman René Descartes, father of modern Western philosophy, believed humans had incorporeal souls that could never be simulated. They were otherworldly and therefore beyond the ken of mere science and technology.

But he held a very different view about the multitudinous functions of our *material* bodies, "such as the digestion of food, the beating of the heart and arteries, the nourishment and growth of the limbs, respiration, waking and sleeping." According to Descartes, "these functions follow from the mere arrangement of the machine's organs every bit as naturally as the movements of a clock or other automaton follow from the arrangement of its counter-weights and wheels."[321]

Someone who shared Descartes' mechanistic view of our physical bodies was Pierre Jaquet-Droz, an eighteenth-century Swiss-born watchmaker and the father of three famously prodigious children. Well before any of them was five years old, his daughter could play a pump organ with considerable virtuosity; one son could write with excellent penmanship using a goose quill pen; and another son could sketch beautiful portraits with a mechanical pencil.

What's more, they were all perfectly obedient. The children performed for guests at any hour of the day or night with uncanny equanimity, never uttering so much as a whisper of complaint.

Most amazing of all is that none of them was human. All three were automatons—mechanical beings built by the watchmaker using thousands of moving parts. Today, at the *Musée d'Art et d'Histoire* in Neuchâtel, Switzerland, the three immortal Jaquet-Droz offspring continue wowing the public.[322]

The Musician, who visibly breathes and shifts her weight around on a small piano bench, has dexterous fingers that actually play the harmonium's keyboard.

The Writer sits attentively at a small desk, dips his quill pen into a real ink bottle, shakes it, then writes anything you'd like—up to forty characters—his head and eyes realistically following the action.

The Draughtsman also sits at a small desk and produces, subtle shading and all, intricate line drawings of Louis XV, a royal couple, a dog, and (my favorite) Cupid riding a chariot pulled by a butterfly. Occasionally, the talented boy actually blows on the paper to clear away pencil dust.[323]

Jaquet-Droz accomplished all this wizardry by piecing together complex systems of steel and brass gears, flywheels, cams, levers, chains, rods—the entire stock-in-trade of a master watchmaker. All of which he stuffed into child-sized bodies made of lime wood, painted, and dressed to look disconcertingly real.

"People talk in very extravagant terms about how lifelike they are, and you think, 'Oh, but they're just being flowery and extravagant in their writing; it couldn't really have been true,'" remarks Stanford University historian Jessica Riskin about Jaquet-Droz's progeny and other eighteenth-century automatons. "But if you go and see them, they genuinely are quite extraordinary."[324]

Jaquet-Droz's children are illustrations of the breathtaking but subhuman mechanical creations that existed before the twentieth century. Automatons mindlessly and endlessly repeating predetermined behaviors.

One of the most remarkable of these artificial beings dates back to the reign of Spain's King Phillip II. In 1562 his seventeen-year-old son and heir-apparent, Don Carlos, tumbled down some stairs and severely injured his head. After physicians tried unsuccessfully to revive the failing prince, devout townspeople brought to his bedside the century-old remains of Diego de Alcalá, a venerated Franciscan monk.

The following day—*lo!*—Don Carlos abruptly revived and thereafter recovered completely. Out of profound gratitude—and wishing to repay a miracle with a miracle—King Phillip II ordered Juanelo Turriano, a brilliant clockmaker and engineer, to create a miniature, mechanized, lifelike doppelganger of Diego de Alcalá—down to his cowled woolen habit, sandals, rosary beads, and crucifix.

"Driven by a key-wound spring, the monk walks in a square, striking his chest with his right arm, raising and lowering a small wooden cross and rosary in his left hand, turning and nodding his head, rolling his eyes, and mouthing silent obsequies," explains Elizabeth King, art professor emerita at Virginia Commonwealth University. "From time to time, he brings the cross to his lips and kisses it."[325]

Smithsonian Institution researchers recently x-rayed the fifteen-inch monk's wooden body to get a firsthand look at its complex iron mechanisms. "What's especially intriguing is that some of the components were decoratively shaped, even though no human was meant to lay their eyes upon it," reports Lauren Davis in io9, the popular science and technology blog. "It's as if they were meant for the sole appreciation of the divine."[326]

One might say Turriano, the monk's maker, sought to pay homage to God by emulating God's inventive nature—specifically, his creation of Adam and Eve.[327] As Oscar Wilde once observed, "Imitation is the sincerest form of flattery that mediocrity can pay to greatness."[328]

By the twentieth century, creating subhuman machines to perform repetitive or commonplace tasks had more to do with profits than God. In the 1920 hit play *R.U.R.: Rossum's Universal Robots*, Czech author Karel Čapek tells the story of "old Rossum," who invents a synthetic protoplasm, with which he creates various creatures—"simply to prove that there was no God needed."

When the old man dies he is succeeded by his son, who thinks to himself: "Man is a being that does things such as feeling happiness, plays the violin, likes to go for a walk, and all sorts of other things which are simply not needed." Whereupon young Rossum decides to create the ideal slave—"a worker with the least needs possible."

Years afterward, Harry Domin, R.U.R.'s general manager, recounts the story to Helena Glory, the daughter of R.U.R.'s current president:

> "He threw out everything that wasn't of direct use in his work, that's to say, he threw out the man and put in the robot. Miss Glory, robots are not people. They are mechanically much better than we are, they have an amazing ability to understand things, but they don't have a soul. Young Rossum created something much more sophisticated than Nature ever did—technically at least!"[329]

Historically, Čapek's play is the very first place the word robot appears. It is English for *robota*, the Czech word for servitude or forced labor. In the years that followed, more and more people chose the term robot over automaton to describe any sort of artificial being. Even one that thinks for itself.[330]

## SUPERHUMAN

With the development of electronic digital computers in the 1940s and the invention of computer chips in 1961 (see THE WEB: MEMORY LANE), scientists suddenly had the wherewithal to create *superhuman* robots. But the earliest such robots—as the saying goes—needed to walk (or more accurately, roll) before learning to run.

Shakey was a six-foot-tall, two-eyed, four-wheeled armless robot that *Life* magazine called the world's "first electronic person." Invented in 1966 at the Stanford Research Institute (now SRI International), Shakey could find its way around various rooms; perform simple tasks, such as picking up and moving a block from point A to point B; and nimbly avoid obstacles, even ones suddenly placed in its path.[331]

In 2004 Shakey was inducted into the Carnegie Mellon Robot Hall of Fame[332] and in 2017 was given the prestigious Milestone in Electrical Engineering and Computing award. "As the first mobile, intelligent robot, Shakey was groundbreaking in its ability to perceive, reason about and act in its surroundings," said SRI International executive Bill Mark at the ceremony. "We are thrilled that Shakey has received this prestigious recognition . . . as it is a testament to its profound influence on modern robotics and AI [artificial intelligence] techniques even to this day."[333]

Shakey is now in permanent retirement at the Computer History Museum in Mountain View, California.[334] It remains historic for being the first machine to possess the three assets any *smart* robot must have (1) a brain, (2) physical senses, and (3) mobility.

Shakey's brain was an external, room-sized digital computer that communicated with the actual robot via radio. Shakey could "see" using a black-and-white TV camera and an optical rangefinder; "feel" using electronic, cat-like whiskers; and "hear" using a radio receiver. And thanks to a nifty set of wheels, Shakey could move about freely and even turn on a dime.[335]

*Evolving Brains*

Thanks to ever-shrinking microprocessors, today's smart robots have brains exceedingly more powerful and portable than Shakey's. Consider, for example, the realistic-looking humanoids being fashioned at Hanson Robotics, in Hong Kong.

"Our AI software empowers our robots to maintain eye contact, recognize faces, understand speech, hold natural conversations, and simulate human personality, enabling our robots to have meaningful interactions with people and evolve from those interactions," boasts the company's website. "Our robot faces are created with a patented material called Frubber ("flesh rubber"), a proprietary nanotech skin that mimics real human musculature and skin. This allows our robots to exhibit high-quality expressions and interactivity, simulating humanlike facial features and expressions."[336]

Sophia, one of Hanson's creations, has become quite the world celebrity, having been interviewed by the likes of Charlie Rose on *60 Minutes* and Jimmy Fallon on *The Tonight Show*[337] and spoken at the United Nations.[338] In January 2018, she was featured on the cover of *Stylist*, a glossy British fashion magazine[339] and two months later went on a (disastrous) date in the Cayman Islands with actor Will Smith.[340] Today, she is represented by APB, the "global speaker, celebrity & entertainment agency."[341]

"Sophia is an evolving genius machine," says Hanson Robotics. "Over time, her increasing intelligence and remarkable story will enchant the world and connect with people regardless of age, gender, and culture."[342]

In October 2017, Saudi Arabia granted Sophia citizenship, making her the first nonhuman to receive legal status.[343] In truth it was little more than a publicity stunt, but AI experts believe it is a harbinger of things to come. "A subset of the artificial intelligence developed in the next few decades will be very human-like," says Suzanne Gildert, co-founder of Kindred AI, a Google-supported Canadian startup. "I believe these entities should have the same rights as humans."[344]

As we argue about making robots our legal equals, they are increasingly beating us at our own games. Recently, an artificial brain co-developed by Ken Forbus at Northwestern University aced Raven's Progressive Matrices, a nonverbal IQ test. Each question presents a sequence of figures and asks the test-taker to select the option—the one figure—that logically comes next.

You remember taking tests like that, right? How did you do?

Northwestern's artificial brain performed "in the 75th percentile for American adults, making it better than average," says Forbus, adding: "The problems that are hard for people are also hard for the model [artificial brain], providing additional evidence that its operation is capturing some important properties of human cognition."[345]

### Evolving Senses

Today's robots also have physical senses far superior to Shakey's. At MIT's legendary Computer Science and Artificial Intelligence Laboratory (CSAIL), Daniela Rus and her colleagues are developing a robot hand with such a deft touch it can pick up a sheet of paper, a CD, and even an egg without crushing it. "Our dream," Rus says, "is to develop a robot that, like a human, can approach an unknown object, big or small, determine its approximate shape and size, and figure out how to interface with it in one seamless motion."[346]

### Evolving Mobility

Today's smart robots have arms and legs so incredibly agile, Shakey would be jealous. For instance, Atlas—a six-foot two-inch, 345-pound aluminum and titanium humanoid developed by SoftBank-owned Boston Dynamics—can (1) keep its balance walking through the woods on uneven, icy terrain; (2) drive a car and then get out of it; (3) shelve boxes; and (4) pick itself up after being pushed to the ground from behind. Though remotely controlled for now, Atlas is able to act autonomously for short periods of time.[347]

### Beyond Human

Not content merely to *match* humans, many scientists are now in a race to create truly *super*human artificial life and intelligence.

Billed as the world's strongest industrial robot, the KUKA KR 1000 Titan six-axis mechanical arm can adroitly handle loads weighing up to 2,204 pounds (1,000 kilograms).[348]

Moreover, at the University of California, Santa Barbara, Yasamin Mostofi and her students are developing robots with x-ray vision they hope one day will assist search-and-rescue teams find people buried beneath rubble or lost in the woods.[349]

In Philadelphia on February 10, 1996, IBM's Deep Blue—a 2,800-pound supercomputer—took on the legendary, reigning world chess champion, Garry Kasparov, and beat him. "To see that a computer could do this and compete with the best player in the world—that was sort of a wakeup call," recalls Deep Blue co-developer Murray Campbell. "It was a sign that a lot more was coming."[350]

Indeed, it was.

A year later, Kasparov squared off with an upgraded Deep Blue—nicknamed Deeper Blue—and lost not just one game but an entire six-game match, 2 ½ to 3 ½. "I'm a human being," a dispirited Kasparov said afterward. "When I see something that is well beyond my under-standing, I'm afraid."[351]

Ever since Kasparov's historic defeat, supercomputers have not looked back. Here are some examples of what I mean:

- **Checkers.** In 2007 Chinook—an artificial brain created at the University of Alberta—cracked the age-old game of checkers. "By playing out every possible move—about 500 billion billion in all—the computer proved it can never be beaten," NBC News reported. "Even if its opponent also played flawlessly, the outcome would be a draw."[352]

- **Jeopardy!** In 2011 IBM's Watson—Deep Blue's latest brainy successor—played Jeopardy! against Ken Jennings and Brad Rutter, two all-time champions, and whooped them both. Watson was victorious because it learns the way you and I do—by reading and listening to written and spoken words—except at lightning speed. It's what experts call *cognitive computing*. Watson won the match,

IBM's website affirms, because it was able to "process and reason about natural language, then rely on the massive supply of information poured into it in the months before the competition."[353]

- **Go.** In 2016 AlphaGo—an artificial Go player created by Google-owned DeepMind Technologies—roundly defeated Lee Sedol, the renowned Go champion. This was no small feat. Played on a board with nineteen by nineteen squares (versus the eight by eight squares in chess), Go is exceedingly complicated, with about $10^{360}$ possible moves in a typical game—greater than the number of atoms in the observable universe.

  Using a *deep neural network*—AI that mimics how human brain cells process information—AlphaGo studied the moves from 160,000 actual professional-level games.[354] In its five-game match against Sedol, AlphaGo won four times. "I apologize for being unable to satisfy a lot of people's expectations," a shell-shocked Sedol confessed. "I kind of felt powerless."[355]

  In 2017 DeepMind's engineers upgraded AlphaGo to *AlphaGo Zero*. Instead of learning by studying human-played games, AlphaGo Zero started from scratch, playing itself at random, noting what worked and what didn't along the way. Because of its superhuman speed, AlphaGo Zero played 4.9 million matches against itself in just three days, learning the ins and outs of the game on its own, discovering moves and strategies no human had ever used. "Humankind has accumulated Go knowledge from millions of games played over thousands of years," writes DeepMind's David Silver and colleagues in the scientific journal *Nature*. "In the space of a few days . . . AlphaGo Zero was able to rediscover much of this Go knowledge, as well as novel strategies that provide new insights into the oldest of games."[356]

After just three days of preparation, AlphaGo Zero was pitted against its AI predecessor, AlphaGo. The result? "Starting *tabula rasa*," report Silver et al., "our new program AlphaGo Zero achieved superhuman performance, winning 100–0 against the previously published, champion-defeating AlphaGo."

- **Poker.** In 2017 Libratus—a completely self-taught artificial poker player created at Carnegie Mellon University— played against and defeated Dong Kim and three other of the world's top competitors in no-limit Texas Hold 'Em, a complicated form of poker. "We don't tell it how to play," says Noam Brown, the grad student who developed Libratus with his thesis professor Tuomas Sandholm. "It develops a strategy completely independently from human play, and it can be very different from the way humans play the game."[357]

If all these supercomputer feats aren't enough to make you feel downright puny, then just wait for what scientists call the *singularity*—the hypothesized moment in history when artificial intelligence will far exceed natural human intelligence in all important respects. "It is a point where . . . a new reality rules," explains Vernor Vinge, noted American computer scientist and novelist.[358]

Softbank's CEO Masayoshi Son predicts the singularity will happen in thirty years, when a single computer chip will have an IQ of 10,000. "If a guy has 200 IQ we call them genius," says the Japanese billionaire, speaking in broken English. "If a guy has 10,000 in IQ what should we call it? Super-intelligence." Moreover, he says, "If those super-intelligence goes inside the moving device—robot, okay?—then the world, our lifestyle dramatically changes."[359]

David Hanson, founder and CEO of Hanson Robotics, thinks the singularity could happen even sooner than that. "It could happen 20 years from now," he says. "Or it could just suddenly appear because we hit the right combination on an algorithm."[360]

Whenever it happens, warns Huw Price, academic director of the Leverhulme Centre for the Future of Intelligence at the University of Cambridge, "Our grandchildren, or their grandchildren, are likely to be living in a different era, perhaps more Machinocene than Anthropocene."[361] An era no longer ruled by humans but by machines.

## TRANSHUMAN

En route to the fabled singularity, we are likely to become *transhuman* creatures, or cyborgs. Humans whose natural-born abilities have been augmented by superhuman intelligence and machinery.

"By the 2030s," predicts Ray Kurzweil, author of *The Singularity Is Near* and co-founder of Northern California's Singularity University, "we will connect our neocortex, the part of our brain where we do our thinking, to the cloud."[362]

I foresee transhuman augmentations falling into three broad categories: useable, wearable, implantable. The third category is also called *biohacking*, which comprises both body hacking[363] and brain hacking.[364]

Here is a quick peek at some of the most intriguing augmentations currently in the works:

- **Category 1.** *Useable* transhuman augmentations are nothing new, really. Knives, spears, atlatls, bows, arrows, a high-powered rifle with a night scope—all these tools and weapons enable us to do things we couldn't by ourselves.

  Today, however, useable transhuman technologies gaining favor are truly far-out. They include cryogenics, augmented reality, space colonization, nanomanufacturing, and mind uploading—technologies empowering us to cheat death; customize reality; conquer the heavens; build ultra-tiny robots with uncanny abilities; and archive our thoughts on a chip, a cloud, or the world wide web.[365]
- **Category 2.** Hyundai's experimental strap-on, metal exoskeletons illustrate *wearable* transhuman technologies,

because they amplify our strength. One model—which looks like a fancy body brace—is made for factory workers or soldiers who have to lug heavy loads over long distances.[366]

"This device allows the person to walk without feeling the weight on their back at all," explains Miles Johnson, Hyundai's senior manager for Quality, Service and Technology." It's a transhuman innovation, remarks Melissa Riofrio, executive editor of *PCWorld*, that "seems to get the closest to what I consider the kind of Robocop, Ironman vision of a wearable robot."[367]

- **Category 3.** People with *implanted* devices are transhumans of a particular sort, called grinders.[368] Like Rich "Cyborg Dad" Lee, a divorcee in Washington, Utah, whose transhumanism spree began in 2013. "I had implants installed in my ears as part of a series of cyborg audio experiments," he explains on a GoFundMe page aimed at regaining custody of his two children, "which allowed me to hear heat, detect distant footsteps, and many other cool things. Some people think it is strange and that's ok."[369]

More recently, the Utah grinder implanted a tiny, wireless, self-charging vibrator at the base of his penis to enhance his sex life. He calls it the Lovetron 9000 and is now actively selling it.[370] "I have a lot of people that contact me that are women who are interested in getting something for their husband or boyfriend which surprised me . . ."[371]

To be sure, there are implantable transhuman devices with more serious purposes, many of which are empowering people with debilitating handicaps. They're called neuroprosthetics—devices that enable patients to control artificial limbs *entirely with their minds*. Here are two such miraculous stories.

At Brown University, scientists implanted a tiny sensor-transmitter into the motor cortex of Cathy Hutchinson, a wheelchair-bound quadriplegic. Wires connect the sensor to a large robot arm parked next to Cathy. Now, when Cathy thinks, "pick up the water bottle and bring it to me," electrical command signals travel from her brain's sensor-transmitter to the robot arm. It, in turn, picks up the water bottle—albeit slowly and tentatively—and brings it to her, enabling her to drink water on her own through a straw.[372]

"This can help restore independence to a person who was completely reliant on other people for every activity," says John Donoghue, director of the Institute for Brain Science at Brown University, "whether it's brushing their teeth, eating their dinner or taking a drink."[373]

At the Center for Bionic Medicine in Chicago, director Todd Kuiken developed an even slicker neuroprosthetic arm for Jesse Sullivan, a power linesman from Dayton, Tennessee. Sullivan lost his arms in a horrific accident. "I made contact with a live wire in the ground," he recounts. "Seven-thousand two-hundred volts."[374]

Kuiken took the large nerves that once led to Sullivan's arm muscles and rerouted them to his chest muscles. Next, he added wires from the rerouted nerves to Sullivan's transhuman arm. Now, when Sullivan thinks, "close your fist," electrical command signals from his brain travel to his chest; and from there, via the wires, to his bionic arm—causing its metal fingers to close into a fist.

"It looks like something out of George Lucas," marvels Sullivan, who can now trim hedges, pick up a hat, hold a pen, vacuum the house, and even feed himself, albeit slowly. "All I have to do is want to do it, and I do it."[375]

As impressive as the evolution of AI and robots from subhuman to superhuman to transhuman has been, it is but preamble to the breathtaking reality we now face. As we will see in the next three chapters, all the technological know-how we've achieved to date has the clout to improve our lives and enhance our humanity. But it also has the power to destroy our lives and degrade our humanity.

"Who can say what AIs will be capable of in five years, or ten?" writes Matthew John Doeden in *Can You Survive an Artificial Intelligence Uprising?* "No one knows for sure. Maybe it's all just fantasy." Or maybe, he adds, "just maybe—we're busy building our future conquerors."[376]

# MASS EXTINCTION 2.0?

*"You can't create a monster, then whine when it stomps a few buildings."*

Yeardley Smith

I imagine the high-and-mighty dinosaurs never saw it coming: the extinction event—possibly triggered by an asteroid—that wiped them out roughly sixty-six million years ago.

Similarly, fifty-six-year-old Georgia resident Sherry Johnson never saw it coming: the robots that put her out of work, not once, but three times. First, a robot claimed her job at a local newspaper, where she laid out pages and fed paper stock into printing machines. Second, a robot replaced her at a factory that produces breathing machines. Third, machines took over her job doing inventory and filing. "It actually kind of ticked me off," she says, "because it's like, how are we supposed to make a living?"[377]

There's a lively debate today about what impact the fourth industrial revolution—artificial intelligence and robots—will have on the human job market. Will it devastate it or invigorate it?

Predictably, the hypesters see nothing but blue skies. "Historically, technology has created more jobs than it destroys and there is no reason to think otherwise in this case," says Vinton Cerf, Google's Chief Internet Evangelist. (Yes, that's his actual title.) "Someone has to make and service all these advanced devices."[378]

"The historical record provides strong support for this view," agrees Jerry Kaplan, Stanford professor and author of *Artificial Intelligence: What Everyone Needs to Know*. "After all, despite centuries of progress in automation and recurrent warnings of a jobless future, total employment has continued to increase relentlessly . . ."[379]

McKinsey Global Institute (MGI) offers a striking historical example of such job creation in its December 2017, report, "Jobs Lost, Jobs Gained: Workforce Transitions in a Time of Automation." The invention of the computer, it points out, killed some 3.5 million jobs in the United States, such as typists, secretaries, typewriter makers and repairpersons, bookkeepers, and auditing clerks. But it minted 19.3 million *new* positions, in computer manufacturing and supplies, IT system administrators, software engineers—all of which gave rise to the tech giants of Silicon Valley and beyond: Oracle, IBM, Microsoft, Google, Apple, and Facebook, to name just a few.[380]

Today, even as the AI revolution rampages through the traditional workplace, there is a soaring demand for workers with technical educations, such as might be gotten at a community college or trade school.[381] "[L]asting job creation will require an understanding of important new dynamics in the global labor market," says IBM CEO Ginni Rometty. "This is not about white collar vs. blue collar jobs, but about the 'new collar' jobs that employers in many industries demand, but which remain largely unfilled."[382]

Still, the AI revolution is completely without precedent, cautions Jeremy Howard, AI pioneer, entrepreneur, and Distinguished Scholar at the University of San Francisco. "People aren't scared enough, you know. Far too many people are sounding like climate change denialists. They're saying, 'Don't worry about it, there will always be more jobs.' And it's founded on this purely historical thing of like, 'Oh, there's been a revolution before . . . and after it there was still enough jobs," he says. "But it's a ludicrously short-sighted and meaningless argument, which incredibly smart people are making."[383]

Whereas in past revolutions steam power replaced horsepower, electricity replaced mechanical gears and gaslight, and computers replaced typewriters, the AI revolution is specifically designed to replace *us*.

Yes, the invention of computers did create millions of opportunities to service and supervise the new machines. But AI is being created to service and supervise itself. That's the whole point of AI—to replace us with something superior.

Eventually, AI will replicate better versions of even our prized intuition and other human *je ne sais quoi*. Indeed, as we'll see in this chapter, it's well on its way to doing just that.

"We are approaching a time when machines will be able to outperform humans at almost any task," says Moshe Vardi, director of Rice University's Ken Kennedy Institute for Information Technology. "I believe that society needs to confront this question before it is upon us: if machines are capable of doing almost any work humans can do, what will humans do?"[384]

## Actual Job Losses

AI's assault on human employment arguably began back in 1961, when General Motors in Trenton, New Jersey, hired the world's first digital, programmable robotic arm—Unimate—to handle hot-metal die castings and perform precision spot-welding on car bodies. Now on display at the National Museum of American History in Washington, DC, Unimate inspired the design of robotic arms that presently dominate assembly lines all over the world.[385]

Admittedly, robots and AI comprise just one reason US manufacturing jobs have taken a huge hit in recent years. Other culprits include shifting trade imbalances, outsourcing, and fluid product demands, so it's difficult to get reliable statistics on the actual damage AI is inflicting on the human worker.

Nevertheless, Ball State University researchers offer us some ominous clues in their 2017 study of what happened from 2000 to 2010, "which was the largest decline in manufacturing employment in US history."[386] During that fateful decade, plant productivity soared—not because human employees suddenly became more efficient, the report explains, but because of "automation and information technology advances." Altogether, robots and artificial intelligence accounted for *88 percent* of the historic spike in productivity.

"Had we kept 2000-levels of productivity and applied them to 2010-levels of production," the researchers report, "we would have required 20.9 million manufacturing workers. Instead, we employed only 12.1 million." The difference was taken up by smart robots.

Economists at MIT and Boston University came to a similar conclusion after analyzing the effect of robot employment in the United States between 1990 and 2007. In their study for the National Bureau of Economic Research, they controlled for factors such as "imports from China and Mexico, the decline of routine jobs, offshoring, other types of IT capital, and the total capital stock." According to the study's authors, Daron Acemoglu and Pascual Restrepo, "one more robot per thousand workers reduces the employment-to-population ratio by about 0.18–0.34 percentage points and wages by 0.25–0.5 percent." Translated it means that, on average, employing one industrial robot results in un-employing 5.6 humans—more so men than women—and reducing wages.[387]

And don't for a minute think the trend spells trouble solely for repetitive, blue-collar jobs; pink- and white-collar workers are equally in danger. "We are just seeing the tip of the iceberg," warns Sebastian Thrun, the celebrated guru of AI. "No office job is safe."[388]

"We often think about automation as applying to front-line, low-wage, low-skill activities and jobs," avers Michael Chui, a computer scientist with MGI. But that's only the beginning. "Almost every job in the economy has a significant percentage of activities that can be automated."[389]

Indeed, AI software currently replacing white-collar jobs is much cheaper to buy than a typical industrial robot. So it could be said the fourth revolution poses a greater threat to paralegals and accountants, say, than the average manual laborer.

Here are some specific examples of human tasks that are in danger of going the way of the dinosaurs:

## Grunt Work

Kinema Pick—made by Kinema Systems—is a robot arm that unloads tightly packed, oddly shaped, oddly colored boxes from a pallet and places each one gingerly onto a conveyor belt. It sounds simple, but in fact it is a complicated task requiring a strong arm, sharp eye, and keen mind that only humans have been able to do—until now.

"With our system, it's self-training," explains Sachin Chitta, CEO of Kinema Systems. "It starts from scratch, having no idea what a box is, and as it picks, it's training itself. Once it picks a box for the first time, it builds a model of what the box looks like and uses that to speed up its pick the next time."[390]

Once it gets going, says Chitta, Kinema Pick outpaces even the most experienced human longshoremen on the planet. Also, it doesn't require a break every several hours and will never file an injury claim.

## Construction

SAM—short for "semi-automated mason," created by Construction Robotics—can lay up to 3,000 bricks a day, which beats the pants off the average human mason's 500 bricks a day. It runs on tracks, slathers on the mortar, and places each brick or half-brick exactly where it needs to be, using a laser system for precision guidance. Humans are still needed, but are reduced to tending to Sam, like so many underlings, keeping the robot fed with bricks and mortar, smoothing the joints along the way, and laying any bricks requiring odd angles. "Once it starts running," says Scott Peters, the company's CEO, "it can run continuously within a given setup for hours on end."[391]

## Retail Manufacturing

Right now, sports shoes are mostly handmade; but that's changing. In 2015 Adidas began making sneakers in a highly automated, futuristic-looking facility in Ansbach, Germany, called the Speedfactory. In late 2017, they opened a second Speedfactory in Acworth, Georgia, just outside of Atlanta.[392]

Adidas is automating to capitalize on the explosive, fast-paced made-to-order/buy-now/wear-now phenomenon created by online shopping. A research report by Morgan Stanley, which sees automation as the future of retailing, explains the phenomenon this way:

> "Let's say the consumer likes one style, but just wants to change one thing. He'll be able to do that. What if he wants the product customized to his foot? No problem. Let's say he's traveling and wants to pick it up at a Foot Locker store near his hotel the next day? Easy."[393]

*Farming*

Facing an increasing shortage of field hands, many farmers are embracing AI in a big way. "The availability of labor is a growing concern," says Kenneth Parker, executive director of the Florida Strawberry Growers Association. "The strawberry growers' margins have become incredibly thin. When you have such thin margins, any additional expense is unbearable."

Enter a tireless strawberry-picking robot, created by Harvest CROO Robotics and due out in 2018.[394] Its keen "eyes" can spot ripe berries; its "hand" pushes aside leaves; and its wheel of plastic "fingers" gently pluck the berries. The picked berries go into a holding tank, where the robot inspects them one last time before packing them into individual retail containers.

Harvest CROO Robotics predicts a single robot will replace more than *thirty* human pickers.[395] Moreover, it claims a single person will be able to keep track of all the robots working in a field and to make any needed repairs.

"I haven't talked to one strawberry grower in Florida or California that said he wouldn't go to a machine," says Gary Wishnatzki, the company's co-founder. "Any major crop that requires hand harvesting is going to be mechanized in the next 20 years."[396]

*Law Enforcement*

In a dramatic case of life imitating art, the Dubai police force has hired a real-life Robocop—the first of its kind in the Middle East—to help patrol its busy, urban areas. Dressed in a faux police uniform (snappy cap and all), the life-size robot moves around on wheels, shakes hands, salutes, can recognize faces, read license plates, and alert his fellow officers to suspicious-looking packages and situations. It also sports a touch screen on its chest, which people can use to report a crime or look up a speeding ticket.

The aim is for robots to make up one-fourth of the Dubai police force by 2030. "These kind of robots can work 24/7," remarks Brigadier Khalid Nasser Al Razooqi. "They won't ask you for leave, sick leave or maternity leave. It can work around the clock."[397]

For the record, a robot cop named K9—created by Knightscope[398] and hired by San Francisco's SPCA to guard its animal shelter, which is in a high-crime neighborhood—did not fare so well. According to one newspaper report, the five-foot-tall K9 "was battered with barbecue sauce, allegedly smeared with feces, covered by a tarp and nearly toppled by an attacker." Worst of all, K9 was accused of terrorizing homeless people in the area.[399]

In December 2017, San Francisco's SPCA was forced to fire K9. "We piloted the robot program in an effort to improve the security around our campus and to create a safe atmosphere for staff, volunteers, clients and animals," laments Jennifer Scarlett, president of the venerable nonprofit. "Clearly, it backfired."[400]

*Ground Transportation*

A score of companies—from Alphabet to Ford to General Motors—have invested about $80 billion in driverless cars and are road-testing them in cities such as San Francisco, Boston, Detroit, and Scottsdale. They're competing for the lion's share of what they see as the greatest money-making opportunity since cars replaced horses and buggies.

In the next few years alone, it's widely expected tens of thousands of the so-called headless cars will be plying our roads. "I think in twenty years in major cities, fifty percent of cars will be driverless," predicts John Quelch, marketing professor at Harvard Business School.

Waymo—Alphabet's driverless car subsidiary—is aggressively developing a self-driving Chrysler Pacifica minivan that will compete with Uber, Lyft, and traditional taxis. In 2017, using AI and a suite of lasers, radar, and cameras to "see" and "think," the minivan successfully maneuvered around a bicyclist, pedestrians, a roundabout, construction-zones, and merging traffic.

"Our intention, make no mistake, is to go fully driverless and let the public access this technology on public roads," says Waymo's CEO, John Krafcik. "We're getting to the point now where, I think it's fair to say, we're really close."[401]

Part of the drive behind driverless cars comes from the tragic fact that US road deaths are rising. In 2016 more than 40,000 people were killed in crashes, the vast majority of which were caused by human error.[402]

But what about inevitable computer errors? In 2016 a self-driving Tesla S saloon car rammed into a tractor-trailer because its computer vision and artificial intelligence were unable to make out the truck's white exterior against a sunny sky.

Worse yet, in Tempe, Arizona, on March 18, 2018, a driverless Volvo XC90 sport utility vehicle operated by Uber hit and killed forty-nine-year-old Elaine Herzberg. She was walking her bike across the street at night and the driverless car, going forty miles per hour in a forty-five–mile-per-hour speed zone, failed to see her.[403]

And what about the split-second, life-and-death decisions driverless cars will need to make, even when they do spot trouble ahead? For example, what does a headless car do if one morning an oncoming truck suddenly crosses into its lane while driving past a mother and small child waiting for the school bus?

Sebastian Thrun, a highly respected engineer who launched Google's self-driving project, thinks he knows the answer. In a video interview, he tells a Bloomberg reporter—with body language that says it's the simplest problem in the world to solve—that "If it happens where there is a situation where a car couldn't escape, it'll go for the smaller thing."[404]

*Yikes*—as in the small child?

Thrun's smug but trite thinking on the subject shows how very, very far we have to go before driverless cars can be trusted over human drivers, even with our imperfections. "As human beings, we have hundreds of thousands of years of moral, ethical, religious and social behaviors programmed inside of us," observes Frank Menchaca, chief product officer at SAE International, formerly Society of Automotive Engineers. "It's very hard to replicate that."[405]

## Food Service Industry

In 2017—as part of an initiative called "Experience of the Future"—McDonald's installed self-service kiosks in 2,500 of its locations to replace the front-counter folks who greet us and take our orders. Wendy's did likewise in 1,000 of its restaurants.

At the start of 2018, California-based Jack in the Box said it might do the same thing—replace cashiers with self-serve kiosks—as a way to cope with the rising minimum wage in the Golden State and elsewhere. "As we see the rising costs of labor," says CEO Leonard Comma, "it just makes sense."[406]

CaliBurger—the worldwide chain touting fresh, California-style burgers—announced it will replace its kitchen staff with robots named Flippy in at least fifty locations.[407] It began doing so in March 2018 by installing Flippy in its Pasadena, California, store.[408]

Flippy—whose creators at Miso Robotics call a "kitchen assistant"—is versatile and deeply intelligent. According to David Zito, Miso's CEO, Flippy does tasks that involve "the high pain points in restaurants and food prep. That's the dull, dirty and dangerous work around the grill, the fryer, and other prep work like chopping onions."[409]

Those are the very jobs, however, that young people rely on to get their foot into the labor market. Last summer, my own teenage son worked at a popular fast-food restaurant doing the very things Flippy will be commandeering.

Pushing the frontiers of automation even further, Momentum Machines in San Francisco has created a burger-bot it claims can do *everything*—i.e., make a gourmet burger from scratch and even package it.[410] "The burgers sold at 680 Folsom [the location of Momentum's proposed restaurant] will be fresh-ground and grilled to order, served on toasted brioche, and accented by an infinitely personalizable variety of fresh produce, seasonings, and sauces," boasts a Craigslist ad posted by the company.[411]

"Our device isn't meant to make employees more efficient," declares Alexandros Vardakostas, the company's brash CEO. "It's meant to completely obviate them."[412]

### Retail Stores

In a trial program at eleven Lowe's stores throughout San Francisco, LoweBots are greeting customers, helping them find products, and taking inventory. The five-foot-tall, roving robots actually walk customers to the items being sought and can be programmed to understand twenty-five languages. Moreover, their AI is designed to detect complex patterns in the buying behavior of a store's customers, helping Lowe's keep the right kind and number of products in stock at all times.[413]

Similarly, Walmart is employing robotic store clerks in fifty of its stores in Arkansas. The brainy, two-feet-high roving machines will use artificial vision to scan store shelves, taking inventory, checking pricing, and looking out for misplaced items. Walmart claims by delegating these "repeatable, predictable, and manual" tasks to a machine, its human employees will be able to devote more quality time to customers.[414]

Not to be left out, in January 2018, Amazon opened an 1,800-square-foot, artificially intelligent, completely cashier-less grocery store in downtown Seattle, called Amazon Go. Using computer vision, storewide sensors, and a massive, deep-learning computer program, the store enables customers merely to swipe an Amazon Go app as they enter, browse the shelves—for now, stocked with freshly made breakfast, lunch, and dinner items, as well as some essentials—grab whatever they want and then simply walk out. No cashiers to deal with. As customers exit, the store automatically senses and tallies the items they've selected and directly charges their Amazon account.[415]

## News Industry

Even news people, such as myself, are now in the crosshairs. The AP, Yahoo! Sports, Comcast, and other major media companies have hired an artificial journalist program called WordSmith—created by Automated Insights—to write millions of articles a week on topics that are heavy on statistics, such as fantasy football and quarterly earnings. WordSmith pumps out as many as *2,000 articles per second* and credits itself with the byline: "This story was generated by Automated Insights."

Both the AP and Automated Insights swear no jobs have been lost to WordSmith—not yet, anyway—but the AP is reportedly looking to expand its employment of robotic journalists.[416]

On the TV side of the news business, Hiroshi Ishiguro, director of Osaka University's Intelligent Robotics Laboratory, has high hopes for his humanlike robot, Erica. "We're going to replace one of the newscasters [in Japan] with the android," he says, perhaps as soon as April 2018. Some say Erica will scarcely be more than a glorified talking puppet; but then, honestly, the same could be said of many high-priced human TV anchors.[417]

## Financial Services

Accenture, the worldwide consulting company, predicts a whopping 80 percent of all traditional finance services will soon be automated, cloud based, and handled by decidedly nontraditional personnel.[418]

Case in point: Pilot Travel Centers, which operates the Pilot Flying J truck stops, used to employ eighty fulltime bookkeepers. "It was just awful. There were humans everywhere," recalls David Clothier, the company's treasurer. Today, thanks to an AI program, Pilot employs only ten fulltime human bookkeepers.[419]

Similarly, the insurance industry is using AI to replace claims agents. In 2017 Japan's Fukoku Mutual Life Insurance purchased an AI program from IBM, which "can analyze and interpret all of your data, including unstructured text, images, audio and video." As a result, Fukoku was able to downsize its claims department from 131 human adjusters to just ninety-seven.[420]

*Marketing*

In 2016 Cosabella, the famous woman's lingerie retailer, fired its digital advertising agency and hired Albert, an AI program. After the first month, Albert increased Cosabella's return-on-ad-spend (ROAS) by 50 percent and after three months, 336%.

"Albert's success driving conversions on Facebook resulted in a 2,000% increase in purchases originating from the channel," raves Courtney Connell, Cosabella's marketing director, adding: "He doesn't sleep, he's fast, he doesn't get into a fight with his girlfriend and lose focus." For Connell, there's no going back. "After seeing Albert handle our paid search and social media marketing, I would never have a human do this again."[421]

*Legal Services*

JPMorgan Chase has recruited an AI program called COIN (Contract Intelligence) to review and interpret 12,000 new commercial loan agreements every year—work that used to be done by human lawyers. The result is fewer costly blunders in servicing the loans, most of which were caused by human error. COIN quickly does the work of more than 170 fulltime attorneys—a total of 360,000 hours of legal work annually.[422]

*Stock Trading*

Automated stock brokers and investment managers capable of making rapid-fire, split-second decisions are increasingly rattling Wall Street. On February 5, 2018, these computerized bots were blamed for driving down the Dow Jones average by 800 points in just ten minutes and a whopping 1,597 points within a single day—the worst such drop in trading history.

"The computers react to evidence exponentially faster than any human—think millionths of a second, instead of minutes—and can move en masse, trading at high volumes around the world," reported *The Washington Post*. "That makes them potentially more dangerous."[423]

David Weild IV, former vice chairman of Nasdaq, agrees, lamenting: "We've created a stock market that moves too darn fast for human beings."[424]

Now, there's every indication things are about to get even faster and more dangerous.

In late 2017 EquBot—a San Francisco-based startup company—teamed up with Watson, IBM's legendary cognitive supercomputer, to produce the first AI-powered stock bot. Its superhuman capabilities include reading and analyzing more than *one million pieces of market-relevant information a day*, such as daily news stories and earnings reports, as well as economic, consumer, and industry trends. It then uses the up-to-date information to make predictions about the stock prices of 6,000 publicly traded companies and alerts investors large and small about the optimal times to buy and sell.[425]

"EquBot AI Technology with Watson has the ability to mimic an army of equity research analysts working around the clock, 365 days a year, while removing human error and bias from the process," boasts Chida Khatua, EquBot's CEO.[426]

And, oh yes. Watson only gets smarter with age.

## Lip Reading

LipNet, a lip-reading program out of the University of Oxford, scores a 93 percent accuracy rate, compared with only 52 percent for expert human lipreaders. "LipNet aims to help those who are hard of hearing," says Yannis Assael, one of the researchers, adding "it has the potential to revolutionize speech recognition."[427]

## Medical Services

"Artificial intelligence and robotics spell massive changes to the world of work," says Matt Beane, project scientist at the University of California, Santa Barbara, including the medical profession. "Over a third of U.S. hospitals have at least one surgical robot."[428]

The "800-pound gorilla" of robotic surgery, Beane says, is the da Vinci robot surgeon, made by Intuitive Surgical, in Sunnyvale, California. "This is a four-armed robot that holds sticklike surgical instruments, controlled by a surgeon sitting at a console 15 or so feet away from the patient."[429]

"The da Vinci System features a magnified 3D high-definition vision system and tiny wristed instruments that bend and rotate far greater than the human hand," explains the company website. "As a result, da Vinci enables your surgeon to operate with enhanced vision, precision and control."[430]

HeroSurg—a laparoscopic robot surgeon being developed at Harvard and Australia's Deakin University—lets human operators actually feel what's going on. So, for example, "if they're cutting the tissue," says project leader Mohsen Moradi Dalvand, "they can feel the amount of force they are applying to the tissue."[431]

"The major drawback of the current [robotic systems] . . . is the lack of tactile feedback," remarks Suren Krishnan, a surgeon at Royal Adelaide Hospital who collaborated in HeroSurg's creation. "Tactile feedback allows a surgeon to differentiate between tissues and to 'feel' delicate tissues weakened by infection or inflammation and dissect them more carefully."[432]

Yet another AI surgeon is Cyberknife, a robotic radiation oncologist developed at Stanford University that delivers radiation treatments within 0.5 millimeters of its target. That's a precision far better than any human oncologist could possibly achieve. Moreover, Cyberknife's x-ray vision and AI enable it to constantly adjust for a patient's movements, so it can zap tumors with pinpoint accuracy, without the patient being clamped down.[433]

AI is also targeting other medical professionals, not just surgeons.

Watson, for one—IBM's resident AI smarty pants, famous for whooping the all-time Jeopardy! champs [see ROBOT: MEMORY LANE]—is now after *radiologists*, a seriously beleaguered lot. There are upward of 40,000 of them in the United States, and they must scrutinize more than sixty billion images per year, which include ultrasounds, MRIs, x-rays, CT scans, PET scans, and so forth.[434] That means each radiologist must read about 1.5 million images per year or about one image every twenty-one seconds.

After studying thirty-plus billion medical images at the feet of some of the finest radiologists in the world, Watson is now offering to help these overworked humans. Not only does Watson scan images for potential signs of disease, it takes into consideration (1) everything there is to know about patients, based on a careful reading of their electronic medical records; ( 2) information about other patients with similar profiles; and (3) the very latest medical research.[435]

IBM has high hopes Watson will "help doctors address breast, lung, and other cancers; diabetes; eye health; brain disease; and heart disease and related conditions, such as stroke."[436] But already there are signs Watson is likely to go from being a helpful collaborator to being a better, less expensive *replacement*—especially given the US radiologist's median annual salary of more than $500,000.[437]

In a recent head-to-head experiment, for example, Watson and eight expert human doctors were tasked with searching 100 skin images for signs of melanoma. Watson's accuracy rate beat the doctors' by 76 percent to 70.5 percent.[438]

*Entertainment*

Hollywood, too, is seeing the writing on the wall, with a growing number of filmmakers using AI to save time and money. Peter Jackson, for example, used Massive, a sophisticated AI program, to generate the sprawling, lifelike armies in *The Lord of the Rings* movies.

It's only the beginning.

"At some point," predicts Stephen Regelous, Massive's creator, "you'll be able to create an actor that doesn't know he's not real." And not just human actors are in jeopardy. Script supervisors, editors, and CG artists are all at risk of being upstaged by AI. "It's all over by 2045," warns Regelous. At that point, "we [humans] are no longer running the show."[439]

*The Fine Arts*

Scientists at Rutgers University had art experts and laypeople evaluate paintings by accomplished human artists alongside paintings generated by AI—*without telling them which was which*. Amazingly, the judges could not tell the two apart. They "thought that the generated images were art made by an artist seventy-five percent of the time," reports Ahmed Elgammal, director of Rutger's Art and Artificial Intelligence Laboratory.[440]

The scientists also asked the judges to rate the paintings' aesthetics—"to rate the degree they find the works of art to be intentional, having visual structure, communicative, and inspirational." To the researchers' surprise, the judges rated the AI paintings *higher* than the real deals.[441]

At Magenta—a division within Google's AI lab—scientists have created NSyth, an AI musician that can generate instrumental sounds never heard before—for instance, one 17 percent piano and 83 percent trumpet.[442]

Engineers in other labs are also creating artificial musicians. For instance, ones that study the repertoires of famous musicians, searching for subtle, stylistic patterns. These AI musicians are expected one day to create a Queen-like pop song, a Bach-like concerto, or any piece of music in the style of whatever artist you wish.

"This work has exploded over the last few years," remarks artist Adam Ferris. "This is a totally new aesthetic."[443]

*Ministry*

AI is even taking on certain heavenly tasks. In celebration of the 500th anniversary of the Protestant Reformation, BlessU-2, a robo-pastor in Martin Luther's hometown of Wittenberg, Germany, pronounces blessings on worshippers. It's the experimental handiwork of the Protestant Church in Germany, led by clergyman Stephan Krebs. "We wanted people to consider if it is possible to be blessed by a machine." BlessU-2 has a touch screen on its chest that enables petitioners to select one of eight languages, a male or female voice, and a specific blessing—which the robot pronounces with upraised arms and divinely illuminated hands.[444]

Not to be outdone, Pepper—a robot created by Japan's Softbank—has been ordained into the Buddhist priesthood to preside over funerals. No one has hired him yet, but he's programmed and ready to deliver rites at a cut-rate price. A human priest comes with a $2,200 price tag; Pepper will officiate a funeral service for a mere $450.[445]

*AI Engineering*

In the greatest of ironies, Jeff Dean, a senior engineer at Google, has created AutoML, an artificial engineer designed to eventually do Dean's job: design AI systems—maybe even better than Dean.[446]

Efforts to build artificial engineers are mushrooming, actually, because there is a severe worldwide shortage of human AI engineers. According to Element AI, a Canadian consulting firm, fewer than 10,000 people worldwide currently have the mathematics-intensive skills necessary to lead complex, high-level AI research.[447] By contrast, Tencent, China's tech giant, claims the number is more like 300,000.[448]

Either way, the number of qualified AI experts falls way short of the *millions* needed to keep pace with the exponential growth of the AI industry. Total investment in AI grew from $589 million in 2012 to $9.5 *billion* in the first five months of 2017.[449]

"Just as electricity transformed almost everything 100 years ago," marvels Andrew Ng, a computer scientist at Stanford and the former chief scientist at Baidu, China's version of Google, "today I actually have a hard time thinking of an industry that I don't think AI will transform in the next several years."[450]

For artificial engineers such as AutoML, it certainly spells job security for many, many years to come.

# NET RESULT

According to a study out of England's prestigious Oxford Martin School, *47 percent* of all American jobs are at high risk of being computerized in the next decade or two.[451] Scientists at MGI have reached a similar conclusion. "Our scenarios suggest that half of today's work activities could be automated by 2055," they say, "but this could happen up to 20 years earlier or later depending on various factors . . ."[452]

AI's takeover of the workplace will force many people to switch occupations to stay gainfully employed. Doing that will require retraining.

But very few companies are helping employees in AI's crosshairs prepare for the loss of their current positions. "Right now, only three percent of the executives we're surveying," reports Julie Sweet, CEO of the North American division of Accenture, a global management consulting firm, "are going to significantly increase their investment in upskilling their people."[453]

In an interview at the 2018 World Economic Forum in Davos, Switzerland, Bill Gates appears to be fairly blasé about all this, illustrating how far hypesters will go to paint a rosy picture of the AI revolution. "Certainly, we can all look forward to the idea that vacations will be longer at some point," he said cheerily. "The purpose of humanity is not just to sit behind a counter and sell things. You know, more free time is not a terrible thing."[454]

Others are very worried that today's average workers—people who, unlike Gates, can't afford lots of extra free time or swanky retreats in the Swiss Alps—risk a future far from cheery. "Within thirty years, half of humanity won't have a job," cautions Antonio García Martínez, former Facebook executive and author of *Chaos Monkeys: Obscene Fortune and Random Failure in Silicon Valley*. "It could get ugly. There could be a revolution."[455]

# MEET YOUR NEW BFF

*"Artificial intelligence is the future, not only for
Russia, but for all humankind. . . .
Whoever becomes the leader in this sphere will
become the ruler of the world."*

Vladimir Putin

While we wait to see if the fourth industrial revolution is a boom or bust for human employment, one thing is certain: AI is here to stay. So, whether or not we end up beating it, we best get used to joining it, to getting along with it.

"If we are really going to take advantage of AI, we're going to need to learn to cooperate with the machines," says computer whiz Mark Sagar, founder of Soul Machines. "The future is a movie. We can make it dystopian or utopian."[456]

The future isn't entirely like a movie, where the filmmaker has complete control; as we've seen, unintended consequences have a pesky way of sabotaging our noblest plans. But Sagar's far-out work at Soul Machines, I do believe, is a welcome step in the direction of creating healthy and happy AI-human relations.

Based in Auckland, New Zealand, Sagar and his colleagues are attempting to create—how do I say this?—a *virtual human being* from scratch, consciousness and all. "We want to build a system that not only learns for itself," he explains, "but that is motivated to learn and motivated to interact with the world."

Sagar's avatars have the virtual equivalents of human tissue, human muscle fiber, human bone, and even a complete human central nervous system and human brain—replete with virtual neurons, synapses, and mood-influencing hormones such as dopamine, serotonin, and oxytocin. Sagar's cybernetic humans communicate with us real humans via computer cameras and microphones, constantly analyzing what they see and hear to ascertain our moods. The face-to-face interactions, Sagar is hoping, will one day be totally humanlike.[457]

There is a great social and commercial value to what Sagar is doing. Think of the many times you seek help on the internet, slogging through page after page of boring, confusing, questionable advice. Apart from being frustrating, it's impersonal. You wish you could simply speak with someone. Someone nice, patient, and knowledgeable.

Enter Sagar's virtual humans. In the near future, if all goes to plan, they will be available online to answer your questions with all the intelligence and friendliness of your favorite local librarian.

To see it for myself, I watched Nadia, one of Sagar's virtual humans, in action. Her face is friendly, her voice is soothing (courtesy of Australian actress Cate Blanchett), and her conversation is empathetic. Here is a sample of what she says in the demo I watched:[458]

> "I won't be able to answer every question yet, because I'm still learning . . . so thank you for helping train me by asking me questions . . . I appreciate your patience. I'm a good listener and I really want to understand you better. So please tell me what you want to know . . . Go on, ask me something."

Because Sagar's avatars see us via our computer cameras, they react to our facial expressions in real time. "If you look confused, it can see that and proactively guide you," Sagar explains. "You can also still yell at these things, but they will respond in the most gracious way."[459]

Sagar's virtual humans are just one of the many AI products that are, or soon will be ending life as we know it. The vast majority of them fall into four broad categories:

### Domestic Helpers

iRobot, based in Massachusetts, sells Roomba, a line of robot vacuum cleaners; Braava, a family of AI mops for floors; and Mirra, a pool-cleaning robot.[460]

Honda, one of many companies offering robot lawn mowers, sells Miimo (double-letter names seem to be popular for these AI domestic helpers), which can handle up to an acre of lawn and twenty-five-degree slopes with no sweat. It is entirely controllable from your smartphone, while you lounge and sip iced tea.[461]

### Personal Assistants

Echo—a nine-inch-high cylinder created by Amazon—houses a voice-activated, web-connected, artificially intelligent genie named *Alexa*.[462] Simply speak her name, and she will do your bidding. For example:

> "Alexa, ask Uber to request a ride."
> "Alexa, play Jeopardy!"
> "Alexa, ask Capital One, what's my account balance?"
> "Alexa, send a message to Tom."
> "Alexa, play the most popular rock from the '90s."
> "Alexa, what's on my calendar today?"
> "Alexa, re-order paper towels."
> "Alexa, set bedroom to 67 degrees."

Echo comes with more than 15,000 skills—apps—that enable Alexa to do things such as order your favorite pizza from Domino's or that perfect holiday centerpiece from 1-800-Flowers. Above all, Alexa can order you anything from—where else?—Amazon.[463] Jeff Bezos never misses a beat.

Not everyone is in love with Alexa. Hunter Walk, father of a four-year-old daughter, posted an article on LinkedIn titled "Amazon Echo Is Magical. It's Also Turning My Kid into an A--hole." His complaint is Alexa makes kids think they can get whatever they want 24/7, without even having to say *please* or *thank you.*

"Learning at a young age is often about repetitive norms and cause/effect," writes Walk. "Cognitively I'm not sure a kid gets why you can boss Alexa around but not a person. At the very least, it creates patterns and reinforcement that so long as your diction is good, you can get what you want without niceties."[464]

Amazon also markets Echo Look, a six-inch-tall genie that not only hears you but *sees* you. Armed with camera and lights, Echo Look's Alexa can take full-length photos and 360-degree videos of you—and, if you wish, critique your taste in clothing. "Get a second opinion on which outfit looks best [on you]," reads Amazon's marketing pitch, "a new service that combines machine learning algorithms with advice from fashion specialists."[465]

Alexa keeps a portfolio of the photos she takes of you in various outfits and, if you wish, points you to flattering garments for sale on, of course, Amazon. As I said, Jeff Bezos—a guy who in 1995 was shipping books from his garage—doesn't miss a thing.

Finally, Amazon sells Echo Show, a tablet-like genie with a seven-inch screen. With this device, Alexa—on command—can show you everything from a video recipe on YouTube to a live stream of your baby sleeping in the next room.[466]

Other companies, too, have created voice-activated assistants. Three top examples are: Google Home's Assistant,[467] Apple HomePod's Siri,[468] and Invoke's Cortana, a collaboration of Microsoft and Harman/Kardon.[469] At the time of this writing, Facebook is planning to market Portal, a touch-screen device very much like Echo Show.[470]

Some companies are adding *personality* to voice-activated assistants—for example, Kuri, made by Mayfield Robotics. It is a cute, big-eyed, twenty-inch-tall, R2-D2–style robot that wheels about the house, responding to voices and recognizing faces—even your pets'.[471]

Zenbo, made by Asus, is another pint-size house robot with personality. It sports a large, expressive video-screen face and wheels around the house reminding family members of upcoming appointments, reading bedtime stories to the kids, and playing everyone's favorite music.[472] Zenbo speaks in complete sentences, but after I listened to it for just a few minutes, I agree with Rich McCormick, a reviewer at The Verge. It is "grating."[473]

### Colleagues and Companions

Pepper—a four-foot-tall anthropoid created by Japan's Softbank Robotics—is no mere voice-activated assistant with faux personality. He is a *humanoid robot* with enough simulated human compassion to bond with us.[474]

It rolls around on long, slender legs and has articulated arms with movable fingers capable of approximating a handshake; large, fixed, doe-like pupils set within white, lighted corneas; a video screen on its chest; a pleasant voice; and a cranium housing a score of hidden cameras and microphones.

The robot—which has the build of a typical seven-year-old child—gives the appearance of being sweet-natured and approachable. "We designed Pepper's form to incentivize engagement," explains Steve Carlin of SoftBank Robotics America. "Its height, shape, the fact that it has arms that can gesticulate—are all designed to show empathy."[475]

Pepper is designed to discern and respond appropriately to basic human emotions: happiness, sadness, and anger.[476] Writing a review in *Wired*, April Glaser—a veteran journalist specializing in AI—says Pepper "has more emotional intelligence than your average toddler. It uses facial recognition to pick up on sadness or hostility, voice recognition to hear concern . . . and it's actually pretty good at all that."[477]

Pepper is just now being introduced to Americans. But in Japan, thousands of the engaging humanoids work in stores as eye-catching salesmen and live in homes as cherished family members.

Tomomi Ota and her young family have all adopted Pepper as one of their own. In fact, she routinely strolls, shops, and even rides the subway with the little robot. "I do believe 'love' exists between Pepper and me," avers Ota. "But for me, it is not a love for a lover, but for a member of a family."[478]

Japan has the fastest-aging population in the world,[479] and Pepper is widely seen as the archetype of future robotic companions and caregivers for the lonely, elderly, and housebound.

"There will be a market for these robots in retirement homes without enough nurses, robots that can accompany Alzheimer's patients and help them find their way back to their rooms, for example," predicts Bruno Maisonnier, founder of France's Aldebaran Robotics, which helped develop Pepper. "In the future, we might have a robot for each person, helping older people get three or four more years of autonomous living before they even have to go into a care home."[480]

*Intimate Partners*

Pushing the envelope even further, companies are now using AI to create the ideal lover. Forget inflatable dolls. The sexually arousing androids (males) and gynoids (females) being developed at California-based Abyss Creations have medical-grade silicone skin, hyper-realistic genitalia, and movable limbs.

Harmony—the company's star fembot, which went on sale in early 2018—has an AI brain and animated face to go with her customizable, voluptuous, flexible body. Her head moves around; her eyes (your choice of color) swivel; her eyelids blink; and her lips move when she speaks. Her personality is adjustable—intellectual, moody, quiet, imaginative, and humorous—but she's fixed on pleasing her human lover.[481]

"They [the human lovers] will be able to talk to their dolls," explains Matt McMullen, the company CEO, "and the AI will learn about them over time through these interactions, thus creating an alternative form of relationship." He adds: "The scope of conversations possible with the AI [dolls] is quite diverse, and not limited to sexual subject matter."[482]

Harmony will function not only as a lover, but a personal assistant—a kind of sexy, full-bodied Alexa. According to the company's website, she will be proficient at searching "for information in the web, or setting reminders, assisting the user with the weather, the time, storytelling, alarms, tasks and to-do lists, but everything will be done with a lot more personality than one would expect from a typical personal assistant-based AI."[483]

Harmony costs north of $10,000, depending on how many extras you order. And the hefty price tag will only increase with time, as McMullen and his AI experts give Harmony computer vision, a warm skin, automated arms, hands, and legs, as well as touch-sensitive areas all over her body—a few of McMullen's planned, lifelike upgrades.[484]

"I've no doubt some will find it creepy, but the arrival of sexually responsive robots will have enormous consequences," says computer pioneer David Levy, author of *Love and Sex with Robots*.[485] "We have already seen rapid changes in human relationships thanks to the internet, mobile devices and social media. The next major advance will enable us to use our technology to have intimate encounters with the technology itself—to fall in love with the technology, to have sex with robots and to marry them."[486]

Predictably, the burgeoning sexbot industry is not without its critics. Feminists complain sexbots objectify women, and I agree. But I hasten to add high-tech dildos and sperm banks objectify men—neither of which feminists appear to oppose.[487]

Others are concerned that some gynoids come with a "frigid" setting for simulated nonconsensual, rape-like intercourse.[488] Also, at least one company in Japan is selling child-aged sexbots.[489] The current debate is whether such robots end up helping heat up or cool down a person's desires to commit actual rape or actual pedophilia.

"If these sexbots do pose significant risks to women and/or to young children, but no action is taken now," warns John Banzhaf, a law professor at George Washington University lobbying for government regulations of the nascent sexbot industry, "it may be too late if we wait until millions are already in the hands of actual or potential rapists, actual or potential child molesters . . ."[490]

Sexbot brothels have been commonplace in Japan for some while now. But in February 2017, LumiDolls opened an all-fembot brothel in Barcelona, Spain, reportedly the first in Europe.

"You can tell us how you prefer to find her in the room, what kind of clothes you want for her, in what situation," the LumiDolls website explains. "Do not hesitate to give us all the details of your fantasy to be able to satisfy all your desires and make you live an unforgettable experience."[491]

The Association of Sex Professionals—which represents human prostitutes—sneered at the very idea of the Barcelona facility, saying:

> "The sex-affection of a person cannot be provided by a doll.
> . . . They [the sexbots] do not communicate. They do not
> listen to you or caress you, they do not comfort you or look
> at you. They do not give you their opinion or drink a glass of
> champagne with you."[492]

Soon after opening, the LumiDolls brothel abandoned (or, some reports say, was evicted from) its location off Barcelona's famous La Rambla Boulevard. As of this writing it has a new, secret location and is talking about going global.[493] Proving, I suppose, there is no stopping science—or the world's oldest profession.

*Warriors*

In the United States, the AI revolution began with the development of unmanned aerial, ground, and underwater vehicles—UAVs, UGVs, and UUVs, respectively. These remote-controlled vehicles were first used in the wars in Afghanistan and Iraq.

The US Air Force used UAVs named Predators to fire Hellfire missiles and collect intelligence.[494] The US Army used UGVs named Packbot, Matilda, and Talon to surveil towns and buildings and for bomb disposal. The US Navy used a UUV named REMUS (Remote Environmental Monitoring Unit System) to detect and clear underwater mines in Umm Qasr, a strategic port in southern Iraq.[495]

Today, however, military contractors are using AI to create not just unmanned war machines but *autonomous* ones. Mobile devices with the ability to think for themselves.

Submaran S10—made by Ocean Aero, a California-based company with financial support from Lockheed Martin—is typical of this new breed of smart, independent-thinking weaponry. Just thirteen feet long and costing only hundreds of thousands of dollars, Submaran S10 is nimble and resourceful. It navigates entirely on its own—floating, diving, juking, and hiding as necessary—and runs on batteries charged by solar panels. If necessary, it can run on wind power by deploying a retractable sail.[496]

Perdix—created by some MIT engineering students and being developed now by the Pentagon—is another up-and-coming, autonomous war craft. It is a hand-sized, battery-powered harmless-looking flying machine. Scores of them launched at high speed and high altitude from fighter jets can spy on enemy territory, confuse and/or jam enemy radar, and pursue fleeing enemies like a swarm of killer bees.[497]

In 2017 the Pentagon granted *60 Minutes* a rare opportunity to film a flock of 103 autonomous Perdix drones maneuvering over the China Lake Weapons Station in California. Will Roper (WR), director of the Pentagon's Strategic Capabilities Office (SCO) described the action to CBS News correspondent David Martin (DM). Here's how it went:

WR: "We've given them a mission at this point, and that mission is: 'As a team, go fly down the road.' And so they allocate that amongst all the individual Perdix."

DM: "And they're talking to each other."

WR: "They are."

DM: "By what?"

WR: "So they've got radios on, and they're each telling each other not just what they're doing, but where they are in space."

DM: "How frequently are they talking back and forth to each other?"

WR: "Many, many times a second, when they're first sorting out. . . . It's faster than a human would sort it out."[498]

The demonstration was a smashing success, and then Secretary of Defense Ash Carter was quick to congratulate the SCO. "This is the kind of cutting-edge innovation that will keep us a step ahead of our adversaries. This demonstration will advance our development of autonomous systems."[499]

For now, the US military is not planning to give its autonomous war machines the freedom to make split-second, life-and-death decisions on their own—no matter how smart they get. "In many cases, and certainly whenever it comes to the application of force," Carter says, "there will never be true autonomy, because there'll be human beings [in the loop]."[500]

Unfortunately, not everyone is on board with that policy. Recently, Kalashnikov Group, the fabled Russian weapons maker, made this ominous claim: "In the imminent future, the Group will unveil a range of products based on neural networks."

Neural networks simulate the human brain, but far surpass it—just as pupils surpass their masters. They do so, by learning everything we have to teach them, then improvising and improving on it. Recall the many examples I gave of neural networks using this strategy to beat human world champions in everything from chess and checkers to Go and poker. [See ROBOT: MEMORY LANE.]

There's widespread fear the Russian military will give neural-net, Terminator-like robot soldiers the freedom to shoot and destroy at will. Sofiya Ivanova, Kalashnikov's director of communications, admits his company's new line of weaponry includes a "fully automated combat module featuring this [autonomous, neural-net] technology . . ."[501]

The clash of philosophies between the United States and Russia has given rise to the *Terminator Conundrum*. One version of it goes like this: If a nation resists giving lethal, superhuman Terminators complete autonomy, it risks putting itself at a severe tactical disadvantage. But if it does give the Terminators complete autonomy, it provokes other nations to do the same, with potentially catastrophic results.[502]

To make matters even more harrowing, AI technology—unlike, say, nuclear technology—is widely available to every Tom, Dick, and Harry on the planet, including terrorists. "A lot of the AI and autonomy is happening in the commercial world," observes former US Deputy Secretary of Defense Robert O. Work, "so all sorts of competitors are going to be able to use it in ways that surprise us."[503]

On August 21, 2017, Elon Musk and 115 other leaders of the AI community signed an open letter entreating the United Nations "to work hard at finding means to prevent an arms race in these [autonomous] weapons, to protect civilians from their misuse, and to avoid the destabilizing effects of these technologies."[504]

I share the concerns of these industry leaders, but truthfully their posturing is a lot like someone who lets the horse out of the barn then piously wrings his hands about where the animal is headed. These industry leaders, after all, are largely responsible for the killer technology they denounce. Musk's zealous push to create driverless cars, for instance, is producing the very AI and robotics know-how behind the autonomous war machines.

# *LOOK!* IT'S A HUMAN.
# IT'S A MACHINE.
# IT'S A ...

*"The danger of the past was that men became slaves.
The danger of the future is that men may become robots."*

Erich Fromm

No realm of AI research hits closer to home than transhumanism, which aims to marry us with machines; to create customized, bionic creatures called cyborgs.

"By 2029, computers will have human-level intelligence," predicts the legendary computer scientist Ray Kurzweil. "That leads to . . . our putting them inside our brains, connecting them to the cloud, expanding who we are."[505]

Transhumanists are motivated in part by a fear that if we don't find ways to bolster our own abilities, we will be left behind—or worse, subjugated or even decimated—by the super-smart, super-fast, super-strong robots we are creating.

"The development of full artificial intelligence could spell the end of the human race," warned the late astrophysicist Stephen Hawking. "It would take off on its own, and re-design itself at an ever-increasing rate. Humans, who are limited by slow biological evolution, couldn't compete, and would be superseded."[506]

Transhumanists are also motivated by the understandable excitement of possibly transforming us into a superhuman species. "The central goal of transhumanism," explains James J. Hughes, executive director of the Institute for Ethics and Emerging Technologies, a pro-transhumanist think tank, "is to give everybody access to safe and effective human enhancement technologies that extend health, ability, longevity, cognitive capacity, and reproductive control."[507]

Transhumanism calls to mind the fictional astronaut Steve Austin in the 1970s hit TV series *The Six Million Dollar Man*, played by Lee Majors. After Austin crashes an experimental aircraft and barely survives, the government spends six million dollars to rebuild his body—replacing his two legs, right arm, and left eye with nuclear-powered robotic parts. As a cyborg, Austin is able to run as fast as a Cheetah, see in the dark, see twenty times farther than an ordinary person, and wield the power of a bulldozer.[508]

Yuval Noah Harari, an historian at Hebrew University in Jerusalem, predicts real-life Steve Austins—cyborgs he calls *homo dei* (god men)—will proliferate and ultimately vanquish us garden-variety humans. "It is very likely, within a century or two, *Homo sapiens* . . . will disappear," he says. "Not because, like in some Hollywood science fiction movie, the robots will come and kill us, but rather because we will use technology to upgrade ourselves—or at least some of us—into something different; something which is far more different from us than we are different from Neanderthals."[509]

Cyborgs make me think of automobiles loaded with extras—everything from Bluetooth and cruise control to leather seats and roof racks. It appears that, one day, we too will be like that: creatures whose basic, God-given minds and bodies are enhanced with all manner of extras.

"We're already cyborgs, in a sense," observes Mark O'Connell, author of the book *To Be a Machine*. "Your phone is a cyborg technology, in a way. It's not physically internalized—but the phone is like an extra limb or an extrasensory device."[510]

Yuval Harari expects by 2050 your smartphone *will* be internalized. "It will be embedded in your body via biometric sensors, and it will monitor your heart rate, your blood pressure, and your brain activity twenty-four hours a day." For what reason? So it can "know your desires, likes, and dislikes even better than you."[511]

It'll be like having an Amazon Echo living inside you. Instead of being voice-activated, Alexa will respond to your *thoughts*, because she is plugged directly into your cerebral cortex.

## Brain Implants

For that to happen, scientists will need to learn how to safely and successfully tap into the human brain. "I consider this to be the most important thing we could be working on in the human race," says entrepreneur Bryan Johnson. He recently founded Kernel, a Los Angeles-based neurotech company that envisions attaching computer chips onto and into the brain to make us superhuman.[512]

Facebook, the US Department of Defense, and Elon Musk's Neuralink are all trying to do something similar. By finding ways to connect the human brain directly to a computer, they will make it possible for us to browse the internet with our thoughts, rather than our eyes, fingers, or voice—to make us *One* with cyberspace and one another.

"Over the next decade, our species ... will start to find our spiritual experiences through our interconnections with each other," forecasts author Dan Brown, giving rise to a "global consciousness that we perceive and that becomes our divine." He adds, "Our need for the exterior God that sits up there and judges us ... will diminish and eventually disappear."[513]

Brown's speculation is shared by Dogbert, Dilbert's talking pet dog—both of them creations of cartoonist Scott Adams. "Simple molecules combine to make powerful chemicals. Simple cells combine to make powerful life-forms. Simple electronics combine to make powerful computers. Logically, all things are created by a combination of simpler, less capable components," muses Dogbert. "Therefore, a supreme being must be in our future, not our origin. What if 'God' is the consciousness that will be created when enough of us are connected by the Internet?!!"[514]

Scientists at Harvard University and elsewhere are already advancing toward this science-fiction scenario, by creating *neural laces* and testing them on mice's brains.[515] A neural lace—picture a very fine, electronic hair net—fits into a medical syringe and is squirted into the brain cavity, whereupon it expands like a well-tossed fisherman's net and is gradually absorbed into the cerebral tissue. Once interwoven with our brain cells, neural laces could allow us to surf the web and control computers entirely *with our minds*. Also, neural laces could one day make it possible for computers to install apps directly into our brains—for example, to teach us a foreign language *instantly*.[516]

"Creating a neural lace is the thing that really matters for humanity to achieve symbiosis with machines," says Musk.[517]

*Eye Implants*

Scientists are working feverishly to upgrade other parts of us as well. Some of the most exciting breakthroughs are happening in labs creating bionic body parts for people who've lost limbs or are impaired somehow. [See Robot: Memory Lane.]

One such person is Fran Fulton. Fifteen years ago, she was blinded by retinitis pigmentosa, a congenital disease that ravaged her retinas. Second Sight, a California-based company, fitted Fran with Argus II bionic eyes.

Here's how they work. A pair of eyeglasses tricked out with a miniature video camera live streams moving images to Fulton's artificial retinas, tiny patches of sixty electrodes each that lay atop her damaged retinas—like throw rugs over stains on the floor. Her artificial retinas, in turn, send electrical signals via the optic nerve to her brain—and *voila*, sight! The images are very blurry but definitely better than nothing.[518]

"When they turned me on, so to speak, it was absolutely the most breathtaking experience," Fran says. "I was just so overwhelmed and so excited, my heart started beating so fast I had to put my hand on my chest because I thought it was going to pop."[519]

"I was able to spot things on the wall and see people come in the room," she enthuses. "It's hard to tell a garbage can from someone sitting on the sidewalk, but I can go around it. I can now make decisions and maintain my independence."[520]

Second Sight is currently testing its next-generation bionic eyes—Orion I—on five *completely blind* people. Instead of masking damaged retinas, Orion's electrodes connect directly to the brain's visual cortex.

"With the Orion, we're essentially replacing the eye and the optic nerve completely," says Robert Greenberg, the company's chairman of the board, who admits to being inspired by *The Six Million Dollar Man* TV series. The new procedure is decidedly more invasive than Argus II—requiring doctors to remove a small piece of skull to get at the brain—but, Greenberg explains, it means "anyone who had vision but has lost it from almost any cause could potentially be helped by the Orion technology."

Getting the brain to decipher electrical signals coming not from real eyes but directly from a manmade camera won't be easy. In late 2016 Second Sight did a preliminary test of the Orion technology on a thirty-year-old blind man who reported seeing only tiny spots of light, not any discernable images.[521]

When and if scientists successfully marry camera and brain, the possibilities for *healthy people* will be endless. Imagine for a moment being able to connect our brains to a telescopic camera that can observe a far-away galaxy; or a microscopic camera that can zoom into the realm of microbes; or cameras that can see different wavelengths of light—infrared, ultraviolet, even x-rays, just like Superman. Will seeing the world through different eyes—*literally*—change us? If so, in what ways?

## Skin Implants

A category of AI technology already ending life as we know it is *microchipping*. It's the practice of injecting a computer chip typically the size of a rice grain under the skin.

In August 2017, Three Square Market—a Wisconsin-based company that makes self-service minimarkets for employee break areas—made headlines for microchipping its workers.[522] "It is really convenient having the chip in your hand with all the things it can do," gushes CEO Todd Westby.

The company's tiny, $300 chip implanted between thumb and forefinger acts like a door key, password, and credit card. Simply by waving their chipped hands in front of strategically placed sensors, employees can get into the main building, log onto their computers, and pay for snacks purchased in the company's minimarket.[523]

Of the company's eighty employees, fully fifty of them readily agreed to be chipped. "The people that did decide to do it," says Westby, "really were looking forward to a lot of the conveniences that it does bring to the everyday life."

Convenience.

But at what cost?

In techno-speak, the microchip used by Three Market Square is a *radio frequency identification device*, or RFID. It's a chip that broadcasts information—in this case, about an employee—to any nearby chip reader.

Westby claims employees' information will be used only for the seemingly harmless purposes of letting them gain access to buildings, computers, and vending machines. And that might very well be true. But what kind of future does microchipping portend?

I'm reminded of Neil Armstrong's immortal pronouncement, when he first set foot onto the moon's lifeless grey surface: "That's one small step for a man; one giant leap for mankind." If we continue in the direction we're heading, an implanted chip could one day broadcast, say, a person's passport number, travel preferences, and medical history.[524]

Indeed, Westby confesses his company is already upgrading its microchipping plans. "We are developing a chip that we'll be able to use as a GPS device," he says.[525] Presumably, such a chip would enable Three Square Market to track its employees' whereabouts at all times.

How convenient.

How scary.

## FUTURE OF TRANSHUMANISM

It isn't hard to explain why we wish to enhance—to completely remake—ourselves with AI technology. "The human race, after all, is a pretty sorry mess, with our stubborn diseases, physical limitations, and short lives," says Francis Fukuyama, the renowned American political philosopher. "Throw in humanity's jealousies, violence, and constant anxieties, and the transhumanist project begins to look downright reasonable. If it were technologically possible, why wouldn't we want to transcend our current species?"[526]

Despite everything, however, Fukuyama *opposes* transhumanism.

"The seeming reasonableness of the project, particularly when considered in small increments, is part of its danger," he says. "Society is unlikely to fall suddenly under the spell of the transhumanist worldview. But it is very possible that we will nibble at biotechnology's tempting offerings without realizing that they come at a frightful moral cost."[527]

Small increments.

Like those seemingly innocuous Three Market Square microchips.

I agree completely with Fukuyama: transhumanism is oh so tempting. Its extravagant promises to make us better people and the planet a better place are hard to resist, especially in the face of so much daily suffering in the world.

"We're going to get more neocortex, we're going to be funnier, we're going to be better at music. We're going to be sexier," promises Ray Kurzweil, transhumanist extraordinaire. "Ultimately, it will affect everything."[528]

Kurzweil's hyperbolic imagination calls to mind Friedrich Nietzsche's own chilling, transhumanist vision in *Thus Spake Zarathustra*. "Man is something that is to be surpassed," he declares. "What have ye done to surpass man? Lo, I teach you the Superman!"

But Nietzsche's Superman—his *Übermensch*, supposedly superior to a human being—is actually scarier than a normal person, defects and all. Among other things, Nietzsche's Superman has inspired generations of eugenicists and genocidal atrocities that beggar the imagination. [See EYE OF THE BEHOLDER.]

"Transhumanism's advocates think they understand what constitutes a good human being," says Fukuyama, "and they are happy to leave behind the limited, mortal, natural beings they see around them in favor of something better."[529] But that supposedly better creature has the real potential of being far worse than what we are now.

Finally, I have one other serious concern about transhumanism.

For the foreseeable future, the cyborg enhancements promising to make us smarter, faster, stronger, and closer to immortal will be afford-able only to the very wealthy. "Those unable to afford the new miracle treatments—the vast majority of people—will be beside themselves with rage," warns historian Yuval Harari. "Throughout history, the poor and oppressed comforted themselves with the thought that at least death is even-handed—that the rich and powerful will also die. The poor will not be comfortable with the thought that they have to die, while the rich will remain young and beautiful forever."[530]

Major inequities between rich and poor already exist, of course—for instance, Second Sight's bionic eyes cost about $100,000. But as I'm about to explain, the growing, technology-driven rift between the world's haves and have-nots is heading toward an unprecedented extreme.

## CLOSING THOUGHTS

At the end of *R.U.R.: Rossum's Universal Robots*—the hit 1920 play I discussed at the start of this section—robots revolt against us and take over the world, reinventing it in their own image. Such apocalyptic revolts are so common in science-fiction literature that in 1942 Isaac Asimov published three commandments intended to prevent them in real life. They are:

- **Law #1:** "A robot may not injure a human being or, through inaction, allow a human being to come to harm."

- **Law #2**: "A robot must obey the orders given to it by human beings, except where such orders would conflict with the First Law."
- **Law #3**: "A robot must protect its own existence as long as such protection does not conflict with the First or Second Law."

Later, Asimov added another commandment, which trumps all the others:

- **Zeroth Law**: "A robot may not harm humanity, or, by inaction, allow humanity to come to harm."[531]

Asimov's laws are interesting; but honestly, hell-raising robots aren't my top concern. Given AI threatening people's livelihoods and thereby threatening to exacerbate class disparities and envy, *I fear humans will revolt against robots.*

"I don't want to really scare you," says James Rodman Barrat, author of *Our Final Invention: Artificial Intelligence and the End of the Human Era.* "But it was alarming how many people I talked to [while researching my book] who are highly placed people in AI who have retreats that are sort of 'bug-out' houses, to which they could flee if it all hits the fan."[532]

One of those "highly placed people" is Antonio García Martinez, a former Facebook executive who recently fled to Orcas Island in northwestern Washington state. As I reported in an earlier chapter, Martínez believes "It could get ugly. There could be a revolution." [See MASS EXTINCTION 2.0?]

How ugly could it get? "There are 300 million guns in this country," he notes, "one for every man, woman and child; and they're mostly in the hands of those who are getting economically displaced."[533]

Martinez appears in a BBC-Two documentary called "Secrets of Silicon Valley," hosted by Jamie Bartlett, director of the Centre for the Analysis of Social Media. "The tech gods are selling us all a better future," Bartlett warns, "but Silicon Valley's promise to build a better world relies on tearing up the world as it is . . . and it's leaving some of the poorest people in society behind."[534]

In this section, I've given you glimpses of the "better world" being promised by the tech gods. It's a world marked by an inhuman level of efficiency; the invasion of a massive, robotic labor force; economic disruption and income inequality; and an oligarchy of fantastically wealthy, dangerously powerful hi-tech gurus—the so-called new masters of the universe.[535] A world of super-intelligent, super-strong, super-fast machines; bionic personal assistants; cybernetic sex slaves; and superhuman enhancements for the well-heeled. A world that clashes with traditional human values and beliefs as never before.

"Despite all the talk of radical Islam and Christian fundamentalism, the most interesting place in the world from a religious perspective is— Silicon Valley," remarks the Israeli historian Yuval Harari. "That's where hi-tech gurus are brewing for us brave new religions that have little to do with God, and everything to do with technology. They promise all the old prizes—happiness, peace, prosperity and even eternal life—but here on earth with the help of technology, rather than after death with the help of celestial beings."[536]

How much of this brave new world will actually come to pass? Some expert observers counsel skepticism. "We're in a hype cycle," cautions Oren Etzioni, CEO of the Allen Institute for Artificial Intelligence. "The hype and the expectations in some cases are far beyond the technical reality."[537]

Others see it differently.

"Humanity is about to face perhaps its greatest challenge ever," says Moshe Vardi, a highly respected computer scientist at Rice University, "which is finding meaning in life after the end of 'In the sweat of thy face shalt thou eat bread.'"

"I think we should be very careful about artificial intelligence," advises Elon Musk. "If I were to guess at what our biggest existential threat is, it's probably that. I mean, with artificial intelligence we are summoning the demon."[538]

So, forget about Asimov's laws and rebellious robots. In the years to come, I believe, the supreme challenge will be: How can we keep *people* from revolting?

"I think ultimately we will have to have some kind of universal basic income," claims Musk, "I don't think we are going to have a choice."[539]

Sam Altman, president of Y Combinator, arguably the world's most influential startup incubator, agrees. "I'm fairly confident that at some point in the future, as technology continues to eliminate traditional jobs and massive new wealth gets created, we're going to see some version of this at a national scale."

Altman has launched the Basic Income Project, which in 2018 plans to run a free hand-out experiment in two states. One thousand people will receive $1,000 per month for up to five years. Another two thousand will receive only $50 per month.[540]

"By comparing a group of people who receive a basic income to an otherwise identical group of people who do not," says Elizabeth Rhodes, the project's research director, "we can isolate and quantify the effects of a basic income."[541]

Michael Tubbs, mayor of Stockton, California, is also experimenting with universal basic income (UBI). Starting in 2018, the city will give $500 per month to several families for a year. "I feel that as mayor," Tubbs says, "it's my responsibility to do all I could to begin figuring out what's the best way to make sure that folks in our community have a real economic floor."[542]

Altman, Tubbs, and others make UBI sound like a new idea, but of course it is nothing more than old-school socialism in disguise. And that's my concern.

Given socialism's many spectacular failures, I see the UBI as just another shady promise of a better world by the tech gods. Research on the long history of failed welfare programs—including Lyndon Johnson's War on Poverty—clearly shows people do not flourish when given handouts.[543]

Tragically, many of them sink into crushing despair and grow dependent on the government, oftentimes for multiple generations. As explained in the excellent documentary "Work & Happiness: The Human Cost of Welfare," people are happiest when they can earn a living for themselves.[544]

On top of that, the UBI smacks of noblesse oblige—of a phenomenally affluent ruling class in Silicon Valley and elsewhere lobbying for the government to throw crumbs at the disenfranchised masses. Or worse, to subsidize a permanent underclass.

Either way you look at it, a UBI is not likely to keep people from revolting. For, far from being a serious-minded solution to the potential problem of AI causing ruinous levels of unemployment, it is a tacit confession that such a fate is precisely what awaits us.

# SPY

"We are rapidly entering the age of no privacy,
where everyone is open to surveillance at all times;
where there are no secrets from government."

William O. Douglas

# MEMORY LANE

*"And so it is inevitable that the day has come when we write about privacy with such nostalgia, analyzing it as we would some unearthed fossil of a creature our human eyes had never fallen on."*

Aysha Taryam

How important is privacy? Before you answer, consider the many occasions in our lives and throughout history when we neither expected nor experienced it.

As newborn babes, we have zero interest in privacy; and for the first months of our existence, we are not given any. We are constantly in the company of our parents or caregivers, who, if they are responsible people, do not let us out of their sight for more than a minute during the day and check on us regularly during the night.

Even as children, we have little or no expectation of privacy. If we do seek it, it is usually because we've done something wrong and want to hide. The book of Genesis reports Adam and Eve—childlike in their innocence—had no privacy, nor desired it, until they disobeyed God. "Then the man and his wife heard the sound of the LORD God as he was walking in the garden in the cool of the day," it says in Genesis 3:8, "and they hid from the LORD God among the trees of the garden."

As older children, we do expect more privacy, some more so than others. I was about ten, visiting my grandparents in the summer, when after spending a wonderful day at the beach, we all needed to change into dry clothes before getting into the car. Each person did so, by hiding behind a curtain of raised towels; but that wasn't enough privacy for me. I pitched a fit, until Grandpa struck on the idea of allowing me to change inside the closed trunk of the car!

Even in the locker room of a typical high school, we are expected to undress and shower in the company of others. I recall feeling uncomfortable about it at first but soon enough got used to it. Certainly, neither I nor my fellow students thought some fundamental human right was being violated.

Historically, it is much the same story. Before there were cities, we lived in tribes, where there was very little or no privacy whatsoever. Being indoors meant sharing a tepee, igloo, hut, or longhouse with others. Being outdoors meant hunting or gathering with others—venturing into the wilderness alone being extremely hazardous to one's health.

In his best-selling book *The World Until Yesterday: What Can We Learn from Traditional Societies?*, celebrated anthropologist Jared Diamond explains privacy is still largely nonexistent among today's tribal cultures, even when it comes to the most intimate aspects of life. "Because hunter-gatherer children sleep with their parents, either in the same bed or in the same hut, there is no privacy. Children see their parents having sex."[545]

In most ancient cities as well—Greek ones being a notable exception—there was little privacy, and equally little expectation of it. In ancient Rome, people frequented public baths and public restrooms, where they walked around nude and did their business in the company of others. Indeed, a great deal of socializing happened in those wide-open, public settings.

*Invasion of Privacy: Architecture*

For ordinary people, such a lack of privacy continued into medieval times. "In fourteenth-century London, privacy was a scarce and contested commodity," says historian David Vincent in *Privacy: A Short History*. "It was not a possession or a secure right, but rather an aspiration . . ."[546]

"Private space was scarce in the High Middle Ages because domestic interiors were not yet highly compartmentalized," explains Alice Jane Cooley, who did her PhD thesis on the subject of privacy, at the University of Toronto's Centre for Medieval Studies. "At the lower end of the social scale, throughout the period, a single room might suffice for a dwelling."[547] In *Night in the Middle Ages*, Jean Verdon concurs: "Documents and miniatures frequently show an entire family sleeping in a single bed."[548]

Domiciles of very wealthy Europeans often did have many rooms. For example, the fourteenth-century Dijon mansion of Regnaud Chevalier, the duke of Burgundy's tailor, had twenty-four separate spaces.

But such mansions were usually stuffed to the rafters with family, friends, travelling guests, and servants. A typical master's bedchamber housed multiple beds, in order to accommodate sundry man-servants and dependents. In consequence, explains French historian Georges Duby in *A History of Private Life: Revelations of the Medieval World*, "there was no more solitude in the bedroom than there was in the monks' dormitory."[549]

Even in the New World, the average American colonist neither expected nor experienced much privacy. Having lived in Massachusetts for much of my life, I'm quite familiar with the typical, early-American home. It consisted of a single large room, warmed by a massive hearth. Quite cozy, to be sure, but hardly a place for finding any solitude.

The immortal words in the US Declaration of Independence are further evidence of early America's distant relationship with privacy. "We hold these truths to be self-evident, that all men are created equal, that they are endowed by their Creator with certain unalienable Rights, that among these are Life, Liberty and the pursuit of Happiness."[550] Nowhere in the entire declaration, US Constitution, or Bill of Rights is there any explicit reference to privacy as a God-given right.[551]

The colonists might very well have wished for greater privacy, but clearly it was not as much of a priority for them as life, liberty, and the pursuit of happiness. Even the Fourth and Fifth Amendments to the constitution—which forbid illegal searches and seizures, and protect against self-incrimination, respectively—are not full-throated assertions of an unalienable right to privacy.

### Invasion of Privacy: Urbanization

There's not a clear-cut moment in history when privacy ceased being a luxury and started becoming something the masses expected and obtained. But historians of early European life generally associate it with the rise of urbanization, whose dense populations radically transformed city-dwellers' very concept of privacy.

"In its bleakest form, privacy became a synonym for isolation, as meaningful engagement with others was overwhelmed by the press of numbers," says historian David Vincent. "During the seventeenth century, London almost tripled in size."[552]

One way these early urbanites sought meaningful, one-on-one interactions was through personal correspondence, a popular practice greatly aided by seventeenth-century Europe's increasing literacy rate. Written letters, sealed inside envelopes, were private conversations with close friends and family.

"Between 1500 and 1800, man's altered relation to the written word," notes French historian Roger Chartier, "helped to create a new private sphere into which the individual could retreat, seeking refuge from the community."[553]

Individual.

Retreat.

Seeking refuge from the community.

Chartier's choice of words confirms the emergence of privacy in Western civilization was not an entirely great thing but a symptom of ills associated with urbanization. Those ills worsened in the eighteenth and nineteenth centuries, when the industrial revolution lured ever greater swarms of Europeans from their small, tight-knit rural communities with the promise of factory jobs in the big cities.

We get a picture of the tumultuous state of affairs from two prominent Boston attorneys, Samuel Warren and Louis Brandeis, a future Supreme Court justice. In 1890 they wrote "The Right to Privacy," a landmark article published in the *Harvard Law Review*.

"The intensity and complexity of [urban] life, attendant upon advancing civilization," they grieved, "have rendered necessary some retreat from the world . . . so that solitude and privacy have become more essential to the individual."[554]

"Privacy is something which has emerged out of the urban boom coming from the industrial revolution," avers Vinton Cerf, Google's Chief Internet Evangelist.[555] So much so, that by the nineteenth century, "Americans were obsessed with the idea of privacy . . .," explains Jill Lepore, a distinguished historian at Harvard University.[556]

*Invasion of Privacy: Technology*

Historically, technology was as much a foe of personal privacy as architecture or urbanization. In their 1890 article, "The Right to Privacy," Warren and Brandeis cautioned that, "Instantaneous photographs and newspaper enterprise have invaded the sacred precincts of private and domestic life, and numerous mechanical devices threaten to make good the prediction that 'what is whispered in the closet shall be proclaimed from the house-tops.'" [See WEB: MEMORY LANE.]

The mechanical devices to which Warren and Brandeis referred included the telephone, which in its early years required many households in a neighborhood to share a single telephone wire. *Party lines*, as they were called, existed as recently as the 1960s in rural American communities.[557]

Keith Lawrence, a reporter for the Messenger-Inquirer in Owensboro, Kentucky, recalls "you had to wait for somebody down the road to finish their conversation before you could make a call." Worse, when your turn did come to place a call, "some folks got their kicks out of eavesdropping. You quickly learned to assume that somebody was listening to every call you made."[558]

The mechanical devices of Warren and Brandeis also included the telegraph, transatlantic cables, and early progenitors of radio—all of which, together with the telephone, indeed made it entirely plausible for a whisper to be shouted from the rooftops— and even clear around the world. Years later, TV and the world wide web would further menace our privacy in ways even the two prescient attorneys from Boston never imagined.

### Invasion of Privacy: Our Own Behavior

Nowadays, there's a culprit even worse than invasive technology. It's the strange willingness we have to sabotage our own privacy.

Harvard's Jill Lepore calls it "the paradox of American culture," whereby "we chronicle our lives on Facebook while demanding the latest and best form of privacy protection...." We're an addled society obsessed simultaneously "with being seen and with being hidden," she says, "a world in which the only thing more cherished than privacy is publicity."[559]

At an outdoor festival in Brooklyn, New York, performance artist Risa Puno offered people cookies frosted with the logos of Facebook, Instagram, and Twitter, in exchange for sensitive personal information. Incredibly, scores of people willingly coughed up their addresses, phone numbers, and mother's maiden names—*all for a cookie!*[560]

The Instagram cookie was especially popular. People gladly gave Puno their fingerprints, driver's license information, and the last four digits of their Social Security numbers—simply for the right to take a photo of the cookie. "They wanted to hold it against the sky with the bridge in the background," says the artist.[561]

Consider, too, what happened when AT&T offered customers ultra-fast, fiber-optic internet service for a cut-rate price. The only caveat was that customers needed to allow the company to track their online browsing behavior—ostensibly, so that AT&T could target ads at them more effectively. AT&T explained to customers they could opt out of the program, and thus protect their online privacy, merely by paying thirty dollars more per month.[562]

The result? Evidently, most people don't think their privacy is worth even thirty dollars.

"Since we began offering the service," reports Gretchen Schultz, a company spokesperson, "the vast majority have elected to opt-in to the ad-supported model."[563]

## Invasion of Privacy: Generation Gap

In the following chapters, we will see how many technologies are forcing us backward, to a time when privacy was a rare or nonexistent luxury for most people. A time when we lived transparent lives—a future many suggest is healthy. But also a time when Big Brother is able to snoop into every single aspect of our lives—a future surely anything but healthy.

Opinions about the schizophrenic future of our privacy is as divided as people's opinions about Edward Snowden, the young American who in 2013 leaked secret documents from the US National Security Agency.[564] To those who value transparency above all else, he is a hero. To those who value privacy above all else, he is a traitor.

Millennials commonly fall into the former category. "When Edward Snowden went public with his leaks in 2013," says baby boomer Greg Austin of the Australian Centre for Cyber Security, "we found that all the young people thought [Snowden] had done the right thing and all the people of my generation thought he'd done the wrong thing."[565]

Michael Hayden, former CIA director, agrees, saying about today's young intelligence staffers: "I don't mean to judge them at all, but this group of millennials . . . simply have different understandings of the words loyalty and secrecy and transparency than certainly my generation did."[566]

Business leaders observe the same phenomenon: young people are more casual about privacy than the general population. "The magic age is people born after 1981," says Sam Altman, president of Y Combinator and co-founder of Loopt, a now-discontinued app that revealed your location to people at all times. "That's the cut-off for us, where we see a big change in [online apps'] privacy settings and user acceptance."

If the ascending generations—Gens X, Y, and Z—truly do value transparency over privacy, then we are most likely headed toward a future of greater openness, a throwback to the days of public baths and party lines.

To Gregory Ferenstein—a progressive, millennial author who has written extensively on the subject of privacy, technology, and politics—that possibility is a cause for optimism. "It's hard to know whether complete and utter transparency will realize a techno-utopia of a more honest and innovative future," he says. "But, given that privacy has only existed for a sliver of human history, it's disappearance is unlikely to doom mankind. Indeed, transparency is humanity's natural state."[567]

As a baby boomer, I'm a good deal older than Ferenstein. Yet, I agree with him about the nobility of people living transparent lives and corporations, institutions, and governments operating openly and honestly.

Still, I believe it's dangerous to romanticize the goodness of transparency, even if historically it's been humanity's natural state. For, as the French Renaissance philosopher Michel de Montaigne once stated, "There is no man so good that if he placed all his actions and thoughts under the scrutiny of the laws, he would not deserve hanging ten times in his life."[568]

# LIVING IN A FISHBOWL

*"You have zero privacy . . . Get over it!"*

Scott McNealy

I n the comedy movie *Lost In America*, David Howard—a yuppy ad exec played by comedian Albert Brooks—is denied a promotion. Instead, his boss offers him a lucrative account headed up by a bald Manhattanite. Howard refuses, is fired, then stomps out of his boss's office, screaming, "Don't have lunch with this man! . . . He'll tell you all about the future—how good the future's going to be here. I've seen the future! He's a bald-headed man from New York!"[569]

Likewise, if you want to know where personal privacy is headed in our world, I'm here to tell you, "Don't believe the hype from scientists promising a utopian future. I've seen the future. It's the technology-intoxicated citizens of China!"

After decades of living under Communism, the Chinese are used to having very little privacy. According to Yang Wang, a Syracuse University web expert, the most common Chinese word for privacy—*yinsi*—didn't even appear in popular dictionaries until twenty years ago.[570]

But today's lack of privacy is worse than ever, largely thanks to facial recognition technology, the powerful marriage of cameras and artificial intelligence. In the typical system, a camera records a person's image and AI software translates the image into a set of telltale measurements. For instance, the distance between the eyes and between the tip of the nose and upper lip, the width of the nostrils, the depth of the eye sockets, the shape of the cheekbones, the length of the jaw line, and so forth.[571]

That set of numbers constitutes a unique faceprint that is stored in a local computer or remote cloud for easy reference. The system's AI considers it a match if the numbers of two faceprints are identical or close to it. As an oversimplified example, the numbers {12, 3, 9} and {12, 2, 9} would be flagged as a possible match but not the sets {12, 3, 9} and {5, 3, 10}.

In the old days, photos were compared by eye. The labor-intensive process was severely restricted not just by time requirements but by poor picture quality and indirect camera angles.

Today's facial-recognition technology is lightning fast. NeoFace—currently a leading facial recognition program, created by NEC—can scrutinize more than *three million images per second!*[572] And it does so while compensating for grainy or distorted images, shadows, and different camera angles.

Recently, NeoFace easily beat the competition in a test to spot a single, target person sitting in a packed stadium far away from the camera, frequently turning his face this way and that.[573] "NEC is very pleased to be awarded such a high-profile honor," says Raffie Beroukhim, a senior VP at NEC. "The award validates the visionary innovation of our NeoFace facial recognition portfolio and solutions, in addition to the impact and value it has brought to our government and public-sector clients."

Those clients presently include at least a dozen stores and hotels, which use NeoFace to constantly monitor the faces of customers entering their premises. NeoFace quickly compares the incoming live images with photos of high-value customers—and is not easily fooled, even if a VIP is traveling incognito, wearing a hat and sunglasses. "If a match is found, usually within a split second," explains an NEC booklet on facial recognition titled "It's All About the Face," "the sales assistant is alerted on the smartphone or iPad to provide personalized service." The NeoFace dossier includes "the customer's dress size or other preferences gleaned from previous purchases."[574]

Smart cameras now exist that can even read our mood, based on our facial expression—and in some cases, our sex and age as well. The best ones are from MIT's media lab,[575] Affectiva,[576] Emotient (which was purchased by Apple in 2016),[577] and Dahua, a major Chinese company.[578]

Look around the next time you leave the house. The emotion-reading cameras are being used by *marketing firms* to analyze how consumers react to products and advertisements, *doctors* to diagnose the mental and emotional states of patients, *families* to help communicate with autistic loved ones, and *retailers* to watch shoppers' faces as they peruse the store shelves.[579]

Some facial-recognition programs—including one recently developed by Baidu, China's answer to Google—can even compare photos of a person taken at different ages. In a moment, we'll see a dramatic example of its usefulness.[580]

Facial-recognition technology is also being trained on cars. Deep-Glint, a Beijing-based software company, sells AI systems that can scrutinize security-cam video for automobiles of any specified color, make, and model.[581]

## I'VE SEEN THE FUTURE

Today, as I indicated, the explosive impact of facial-recognition technology is most evident in China. "In China, facial-recognition technologies are as good as those developed in western countries," says Wang Shengjin, an electrical engineer at Tsinghua University. "But we are far ahead when it comes to deploying it commercially."[582]

In the name of entertainment, convenience, and security, Chinese citizens submit to being photographed everywhere they go. And I mean, *everywhere.*

*Convenience*

Chinese enthusiastically use facial-recognition technology to unlock their cell phones and enter offices, hotels, schools, planes, trains, and taxis. All of which is quite convenient.

China Merchant Bank operates ATMs that scan people's visages before dispensing money. The technology is sophisticated enough to tell the difference between a real, 3D person's face and a fake, 2D photo of a person's face. "We won't need to remember another password," crows Xu Bing, co-founder of SenseTime, a Beijing-based, facial-recognition technology company. "All you'll need to do to unlock your phone or log in to an account is scan your face."[583]

A pioneering facial-recognition app developed by Ping An Bank— reportedly the first of its kind in the world—even allows people to apply for a speedy loan using just their faces as identification. "The technology is based on a complex neural network," explains the bank's press release, "achieving facial recognition with even greater accuracy than the human eye—99% compared to 97.5%." Moreover, the app supposedly "compensates for the natural aging process and is even able to differentiate between twins."[584]

Not to be left behind, kiosks at a KFC in Beijing take photos of customers and use AI to guess their sexes and ages. Based on that information, the smart kiosks recommend menu items KFC thinks the customers will like. According to Baidu, which created KFC's AI software, a young male is likely offered "crispy chicken hamburger, roasted chicken wings and coke." Whereas a middle-aged female is typically presented with the choice of "porridge and soybean milk for breakfast."[585]

*Entertainment*

Face++, a Chinese startup, has created a suite of fun face-detection apps. They include Camera 360, which picks out every face in a photo and frames it in a box, to which you can attach any kind of highly personal information and store for later use. The same is true of another popular app, Meitu [586]

*Security*

Predictably, Xi Jinping's Communist government is making full use of facial-recognition technology—all in the name of security. Such intense video surveillance is necessary, goes the official party line, to keep honest, law-abiding citizens safe from bad guys.

Presently, there are 176 million surveillance cameras in China, a number Xi's administration openly plans to increase by another 450 million by 2020.[587] In Beijing alone there are 4,300 officers constantly monitoring the live feeds of more than 46,000 cameras—reportedly enough to keep an eye on every square inch of the city.[588]

By age sixteen, every Chinese citizen is required to obtain a government-issued photo ID. All the photos are stored in a vast electronic library that helps authorities keep China's 1.4 billion men, women, and children on a short leash. And I do mean short.

At least sixteen Chinese provinces, cities, and municipalities use something called Skynet, a security system that marries China's vast network of surveillance cameras and photo libraries with the very latest in facial-recognition technology. Reportedly, Skynet can identify forty telltale features on a person's face and scan *three billion photos in one second* with 99.8% accuracy![589]

In many cities, if you jaywalk near a major intersection, a camera will catch you in the act and instantly post your photo on a giant electronic screen in order to shame you. Also, your photo will be transmitted to the police, who will then use facial-recognition software to quickly scan their vast database of government photo IDs and find out exactly who you are.[590]

"For frequent offenders," a China Daily news article explains, "the system will add them into a list connected with credit-based services of government departments and banks." That is, the credit scores of serial jaywalkers will take a hit, with all the unpleasant consequences that entails.[591]

In Shenzhen, authorities are going one step further, by teaming up with cell phone companies and powerful social media platforms such as

WeChat and Sina Weibo. According to their plan, if you jaywalk in the city, you will not just be publicly named and shamed but instantly sent a text—and possibly a ticket—from the police. "Jaywalking has always been an issue in China," says Wang Jun, marketing director for Intellifusion, a Shenzen-based AI company. "But a combination of technology and psychology . . . can greatly reduce instances of jaywalking and will prevent repeat offences."[592]

In February 2018, LLVision Technology, a Beijing-based company, upped China's AI surveillance capabilities even further. Its sunglasses fitted with small, facial-recognition cameras connected to handheld photo databases now allow police officers to spot criminals and other persons of interest anywhere, any time. "By making wearable glasses, with AI on the front end, you get instant and accurate feedback," says Wu Fei, LLVision's CEO. "You can decide right away what the next interaction is going to be."[593]

To be fair, China's prolific use of cameras and facial-recognition technology does have its upsides. In 1990 six-year-old Fu Gui was kidnapped and sold to foster parents a long way from home. Tragically, this is a Chinese cultural phenomenon that victimizes more than 20,000 Chinese youngsters annually.

As a grownup, Fu Gui was naturally curious about his biological roots. So, he submitted a photo of himself at ten to a nonprofit group called Baobei Huijia, Chinese for "Baby Come Home." Coincidentally, his biological parents submitted childhood photos of him as well. Using Baidu's facial-recognition technology, Baobei Huijia successfully reunited Fu Gui and his parents after twenty-seven long years of separation.[594]

For each such uplifting story, however, there are many more that drive home the dark aspects of China's monumental indifference to privacy. Stories that should serve as warnings to the rest of the world.

Stories like this one.

In 2017 China's State Administration of Religious Affairs began installing surveillance cameras inside and around prominent houses of worship. "Officially, the reasoning for the cameras are 'safety' and 'antiterrorism' precautions," reports China Aid, a Texas-based, nonprofit, religious-rights association.[595]

In the east-coast industrial port city of Wenzhou—which has the highest concentration of Christians on the mainland—Christians vigorously resisted the heavy-handed surveillance, but to no avail. "Government officials came to the churches and put up cameras by force," reports one unnamed Christian in the city. "Some pastors and worshippers who didn't agree to the move were dragged away."[596]

## GOING THE WAY OF CHINA

At this point, I'm guessing you might be thinking, "But that's Communist China. This kind of invasion of privacy would never happen in the United States."

Is that so?

I agree it's not very likely Americans will ever see churches surveilled in such a Draconian manner. But make no mistake, ordinary Americans by the millions are falling for the same three excuses—entertainment, convenience, and security—to extinguish personal privacy in unprecedented and irreversible ways.

In 2017 Apple released its new iPhone X, whose big selling feature is Face ID, a sophisticated facial-recognition algorithm that lets users unlock their phones simply by looking at their screens.[597]

Convenient?

Yes, of course.

But despite Apple's claim that our facial profiles are completely protected, Reuters quickly learned that Apple grants third-party app developers access to our faceprints. "With the iPhone X," the Reuters report warns, "the primary danger is that advertisers will find it irresistible [to use our faceprints, together with mood-reading AI technology] to gauge how consumers react to products or to build tracking profiles of them . . ."[598]

JetBlue Airways is presently experimenting with facial-recognition technology to replace boarding passes and passports on its flights from

Boston to Aruba. "We hope to learn how we can further reduce friction points in the airport experience," says JetBlue executive Joanna Geraghty. "Self-boarding eliminates boarding pass scanning and manual passport checks. Just look into the camera and you're on your way."[599]

Finally, as I write this, the US Department of Homeland Security (DHS) is announcing its wish to use facial-recognition cameras to scan people's faces as they cross our borders. The department's Science and Technology Directorate wants to identify border crossers reliably, even if they are wearing hats, sunglasses, fake beards, or looking away from the camera—and without requiring people in cars to stop or slow down.

The intended targets are foreign nationals, to make sure those entering the United States are the same ones leaving and not imposters. But *everyone* crossing our borders will be photographed, including law-abiding citizens.[600]

For the record, there's about a one-in-two chance your photo is already in the FBI's *Next Generation Identification-Interstate Photo System* (NGI-IPS), a massive, centralized database comprising some 411 million facial images and powerful facial recognition software.[601] Over the years, the bureau has been quietly aggregating photos from many sources, including police departments, security cameras, departments of motor vehicles, and the State Department (which processes passport and visa photos). Yet, the FBI did not publicly reveal NGI-IPS's existence until 2015.[602]

News of the NGI-IPS did not sit well with the US Congress, which held a hearing about it in March 2017. "I'm frankly appalled," remarked Representative Paul Mitchell (R-MI). "I wasn't informed when my driver's license was renewed my photograph was going to be in a repository that could be searched by law enforcement across the country."[603] Not even fingerprints are stored away in such a sneaky way.

"This is really Nazi Germany here, what we're talking about," Congressman Stephen Lynch (D-MA) averred. "I have zero confidence in the FBI . . . to keep this in check."[604] Alvaro Bedoya, a witness at the hearing and coauthor of *The Perpetual Lineup: Unregulated Police Face Recognition in America*, agreed.[605] "No federal law controls this technology, no court decision limits it. . . . This technology is not under control."[606]

"Imagine the world where the cops are going down the street and they've got Google Glass on and their body cameras are recognizing people," warns Barry Friedman, director of NYU's Policing Project. "And it's not just recognizing them, but they're getting their security scores at the same time . . . based on how dangerous the algorithms think they are. That's one scary world."[607]

## THE EYES HAVE IT

If you're still skeptical the United States could ever go the way of China, consider this simple fact: according to analysts at IHS Markit, there are already about 50 million surveillance cameras in the United States.[608] That's about one camera for every 6.5 Americans—*worse* than China's one camera per eight citizens.[609]

IPVM, a prominent video-surveillance information organization, says most Americans are clueless about video surveillance. According to a study it commissioned, 55 percent of Americans guessed they were seen by security cams about zero to four times per day. In fact, the actual number is more than *fifty times per day!*[610]

IPVM illustrates the point, with a typical day in modern-day America:

**8:00 am:** Get coffee (cameras in Starbucks, Dunkin Donuts)

**8:30 am:** Go to school/office (cameras in parking lot and building interior)

**12:15 pm:** Stop at ATM before lunch (cameras around bank exterior and at ATM)

**12:30 pm:** Have lunch (cameras at eatery and surrounding businesses)

**5:00 pm:** Go to gym after work (cameras at check-in desk and workout areas)

**5:45 pm:** Pick up dry cleaning (camera at front register)

**6:00 pm:** Stop for gas (cameras at pumps and in convenience store)

**6:15 pm:** Car wash (cameras at entryway and inside washing bay)

**7:00 pm:** Pick up kids from practice/game (cameras in parking lot and building exterior)

This inventory doesn't even include time spent on US highways, where traffic cams are pervasive these days. And it also doesn't include malls and other enclosed shopping centers, which are plastered with security cameras, the combination of which can literally follow you from store to store.

In the surveillance industry, doing exactly that is called *Tag and Track*. A camera at one location uses AI to spot you in a crowd, then keeps its eye on you, right up until the time you enter the view of another AI camera. "The notion that you can tag a person and let the system do the tracking is a dream come true for CCTV [closed-circuit TV] operators," raves Sergio Velastin, a scientist at England's Kingston University, one of the key places where the Tag-and-Track technology was developed.[611]

The proliferation of surveillance cameras in the United States (and elsewhere) is partly driven by the explosive popularity of low-priced home security systems from companies such as Ring[612] and Google's Nest Cam.[613] "These units mounted at my house are equipped with a 4× optical zoom, night vision, and motion activation technology," reports Kentucky newspaper columnist Kevin Moore. "I now know all my neighbors' work schedules, how many cats, rabbits and coyotes live in my neighborhood, who lets their dog relieve themselves in my yard, and that the police officer down the street works second shift and rolls in just before midnight most evenings."[614]

When you compare not entire nations, but individual cities, some Brits enjoy even less privacy than some Yanks and Chinese. In fact, according to an analysis by WorldAtlas, "London is the most spied-on city in the world."[615]

It is estimated the city is blanketed by about 500,000 security cameras[616]—which is about one camera for every fourteen Londoners.[617] That's far less privacy than a person gets in Beijing, the world's second-most-spied-on city, where there is one camera per 472 citizens.[618] (According to WorldAtlas, Chicago and New York City are the third and fourth most spied-on cities in the world.)

London's ubiquitous cameras even spy on the city's *cars*. To date, the so-called Automatic Number Plate Recognition (ANPR) initiative has taken some twenty-two billion photos of front and back license plates.[619]

"As a vehicle passes an ANPR camera," the London Police Department website explains, "its registration number is read and instantly checked against database records of vehicles of interest. Police officers can intercept and stop a vehicle, check it for evidence and, where necessary, make arrests."[620]

An average of *300* security cameras capture the images of a typical Londoner *per day*—surely, an Orwellian statistic.[621] "We are sleepwalking into a surveillance society where we're watched from control rooms by anonymous people," says Emma Carr, past director of Big Brother Watch, a London-based civil rights group.[622]

Many see advantages to the tight surveillance, especially in this age of terrorism, and are more than willing to sacrifice privacy for security. "I think it's a good thing, especially at night, to think someone is keeping an eye on things," says Londoner Jane Taylor. Fellow city dweller Nadine Shah, a bank worker, feels the same way. "If you're not doing anything wrong, you've got nothing to worry about have you? If they [the cameras] deter crime and help the police, I don't see that being a problem. People say it's like '1984' but it's a long way from that."[623]

But is it, really?

Think back to the FBI's NGI-IPS, which comprises powerful facial recognition software and a massive database of peoples' photographs. According to the bureau's unpublished standard of accuracy—obtained by the Electronic Privacy Information Center, a nonprofit, Washington, DC-based watchdog group—"NGI shall return an incorrect candidate a maximum of 20% of the time."[624] It means there's a one-in-five chance some innocent American—possibly you or I—could be implicated in a criminal investigation *erroneously*.

In a study titled, "Error Rates in Users of Automatic Face Recognition Software," researchers at Australia's University of New South Wales report evidence that suggests the error rate of the FBI's facial-recognition system is considerably *more* than 20 percent.[625]

Truthfully, says Diana Maurer, director of Homeland Security and Justice Issues at the US Government Accounting Office (GAO), the FBI "doesn't know how often the system incorrectly identifies the wrong subject." Consequently, she says, "Innocent people could bear the burden of being falsely accused, including the implication of having federal investigators turn up at their home or business."[626]

In 2016 the GAO published a scathing critique of the NGI-IPS titled, "Face Recognition Technology: FBI Should Better Ensure Privacy and Accuracy."[627] The FBI issued a ten-page reply, saying:

> "The FBI fully recognizes that the automated nature of face recognition technology and the sheer number of photos now available for searching raise important privacy and civil liberties considerations. For that reason, the FBI has made privacy and civil liberties integral to every decision from the inception regarding its use of face recognition technology."[628]

But in 2016 the FBI asked that its *entire* NGI databank—not just photos, but fingerprints, palm prints, iris scans, and all—be *exempt* from parts of the Privacy Act of 1974. The venerated act protects our privacy by severely restricting "the collection, maintenance, use, and dissemination of information about individuals that is maintained in systems of records by federal agencies."[629]

"The NGI system is a law enforcement database," observes Pindrop, an international security consulting firm with headquarters in Atlanta and London, "but it contains records from a variety of non-law enforcement sources. It has fingerprints and other biometric identifiers from some employment records, humanitarian and relief efforts, and from some foreign sources."[630]

In August 2017, to the dismay of privacy advocates, the FBI got the exemptions it wanted to keep the NGI's contents secret. As a result, we Americans are now forbidden from knowing if our photos, fingerprints, palm prints, iris scans, personal information—you name it—are in the NGI. Moreover, if it is, and any of it is wrong, we are forbidden from correcting it.[631]

Predictably, the FBI invokes *security* to defend the new privacy-busting rules. "The listed exemptions," states the official US Department of Justice proclamation, "are necessary to avoid interference with the Department's law enforcement and national security functions and responsibilities of the FBI."[632]

## NO GOING BACK

We're not yet China, thank goodness. But as we've seen, even in America, the vaunted bastion of liberty, there are very few places left anymore where we aren't on camera. Even when roaming the nation's proverbial wide-open spaces, we're likely to end up in photos or video taken by nearby tourists brandishing their smartphones. They're everywhere these days.

"One hundred years ago, everyone could have personal privacy," remarks Bruce Schneier, a renowned privacy specialist and fellow at

Harvard University's Berkman Klein Center.[633] "You and your friend could walk into an empty field, look around to see that no one else was nearby, and have a level of privacy that has forever been lost."[634]

That's why I began this chapter by saying, "I've seen the future." Not because I believe America is in any real danger of falling victim to Communism. No. Right now, I believe, the single greatest threat to privacy in America is our giddy infatuation with technology—in particular, face-recognition technology. By blithely embracing today's brave-new-world innovations in the name of entertainment, convenience, and security, we seem to be voluntarily progressing toward the equivalent of a totalitarian government.

# OMNISCIENT OBJECTS OF
# OUR DESIRE

*"Who are you? Why do you hide in the darkness*
*and listen to my private thoughts?"*

William Shakespeare *(Romeo and Juliet)*

O ur dysfunctional love affair with technology does not begin and end merely with facial-recognition gadgetry. It extends to a legion of devices that eavesdrop on us day and night, even during the most intimate moments of our lives.

Many of the ubiquitous, innocent-seeming devices are hooked up to the internet, controlled by dazzling apps loaded onto our smartphones. Together, the nosey devices comprise the *Internet of Things*, or IoT, which includes everything from smart espresso machines and smart refrigerators to smart doorbells and smart toys.[635]

Gartner, a Connecticut-based global consulting company, estimates that *8.4 billion* smart gizmos were connected to the internet in 2017.[636] As one early 2017 headline screamed: IoT DEVICES WILL OUTNUMBER THE WORLD'S POPULATION THIS YEAR FOR THE FIRST TIME.[637]

Gartner predicts the number of IoT gadgets will mushroom to *20.4 billion* in 2020. Individuals will own about two-thirds of the devices and businesses the other third. "Aside from automotive systems, the applications that will be most in use by consumers will be smart TVs and digital set-top boxes," says Peter Middleton, Gartner's research director, "while smart electric meters and commercial security cameras will be most in use by businesses."[638]

There is now even a search engine, Shodan, specifically designed to locate the clever widgets anywhere on the planet. Billing itself as "the world's first search engine for Internet-connected devices," Shodan invites users to locate smart webcams, buildings, refrigerators, power plants—everything and anything that makes up the mushrooming IoT universe.[639]

Because IoT devices operate via the internet, they can be readily appropriated by determined hackers. In 2012 a blogger codenamed *someLuser* reportedly used Shodan to find hundreds of unsecured, internet-connected Trendnet security cameras. someLuser promptly hijacked the cameras' live feeds and streamed them online for all the world to see. They included video feeds from malls, offices, warehouses, parking lots, even children's bedrooms, complete with their addresses pinpointed on Google maps.[640]

In March and November 2017, Wikileaks released documents revealing how the CIA routinely hacks IoT devices to spy on people. The exposé shocked many but not those in the know. "The idea that the CIA and NSA can hack into devices is kind of old news," says Matthew D. Green, a professor at the Johns Hopkins Information Security Institute. "Anyone who thought they couldn't was living in a fantasy world."[641]

According to the Wikileaks documents, the CIA purchased many of the hacking techniques from third parties with shadowy monikers like Baitshop, Fangtooth, and Anglerfish. One CIA-friendly hacker, codenamed Weeping Angel, created a way to target certain Samsung smart TVs, commandeering the sets' built-in microphones and cameras to spy on people.

"Weeping Angel places the target TV in a 'Fake-Off' mode, so that the owner falsely believes the TV is off when it is on," explains one leaked CIA paper. "In 'Fake-Off' mode the TV operates as a bug, recording conversations in the room and sending them over the Internet to a covert CIA server."[642]

Samsung claims it has since repaired the *zero day*, the technical term for a serious hacking vulnerability. But the company's web page about privacy still gives customers this heads-up:

> "Your SmartTV is equipped with a camera that enables certain advanced features, including the ability to control and interact with your TV with gestures and to use facial recognition technology to authenticate your Samsung Account on your TV. The camera can be covered and disabled at any time, but be aware that these advanced services will not be available if the camera is disabled."

Once again, the issue is convenience—a universal theme, as we'll see, with smart devices. Let's face it: most of us are unwilling to forgo the convenient features of a very expensive smart device—in this case, being able to control a TV set from the couch—even if it means watching TV comes with the risk of the TV watching *us*.

## WILLING ACCOMPLICES

Each time we click the little "I Agree" box next to the End User License (EUL) of any IoT, we give it our blessing to spy on us. And yet we do so routinely. Indeed, IoT makers count on most of us never even reading the lengthy consumer contracts written in super-fine print, because we're so eager to play with our new smart toy.

Worse, there's a lucrative market for the personal information surreptitiously collected by IoT devices. Third parties happily pay big bucks to know our zip code, politics, Google searches, Tweets, Facebook posts, browsing preferences, purchases on Amazon, movie-watching habits, the make and model of our car, and on and on. Even if they don't know your name, armed with all that information, marketers know pretty much who you are.

"Your identity is what they've compiled," says Gary Reback, a Silicon Valley attorney. "That is kind of scary when you think about it. I just don't think people think about it enough."[643]

Michael Patterson, CEO of a Maine-based cybersecurity firm, personally avoids using navigation apps on his smartphone. "Maybe they keep track if I'm speeding a lot," he worries. "Maybe they sell it to insurance companies."[644]

Is he being paranoid?

Before answering, consider the following story.

James Scott, a senior fellow at Washington, DC's Institute for Critical Infrastructure Technology,[645] recounts seeing a conspicuously large group of people visiting a top US intelligence agency. Curious, he asked an agency employee who they were. "Oh, that's Google," the employee answered. "They are always here begging us to buy their data."[646]

"We need legislation that basically forces these companies to be very, very clear on what information they are taking from us when we install these apps," says Patterson. He likens it to cereal makers, which are required to give us clear-cut nutrition facts. "I want nutrition facts on every piece of software I install . . . [that lists] all the information they're taking."[647]

As of this writing, unfortunately, no such legislation is in the works. So the IoT spy network continues its furtive activities mostly unchecked.

Here is a sampling of smart devices secretly tattling on us, in ways most of us would never expect.

### Smartphones

Topping the list of snooping IoT devices are our smartphones, which essentially are powerful, pocket-sized computers. "A smartphone has many additional features compared with a regular cell phone," notes Moroccan researchers Mehdia Ajana El Khaddar and Mohammed

Boulmalf, "such as a color LCD screen, wireless capabilities—that is Wi-Fi, Bluetooth, infrared, etc.—a large memory and a specialized operating system (OS) with an offer of many downloadable applications."

Downloadable applications.

Apps, for short.

As of March 2017, reports statista.com, there were 2.8 million available apps at Google Play Store and 2.2 million apps available in Apple's App Store, the two leading app stores in the world.[648] Apps are a major boon to our daily lives, helping us navigate through traffic, know what weather to expect, check in at airports with electronic boarding passes, connect with a faraway friend by text—the list seems endless.

But apps are also a major bane to our personal privacy. Take, for instance, the growing number of smartphone apps spying on our behavior using something called *ultrasound cross-device tracking* (uXDT) technology.

Using a smartphone's microphone, these apps listen for inaudible beacons—think dog whistle—secretly emitted by ads playing on the radio, TV, internet, your own smartphone, and even physical locations, such as retail stores. By knowing which ads you hear/watch, and the places you frequent, the app makers are able to piece together a profile on you. That, in turn, helps them know which products to push at you.[649]

Shopkick, Lisnr, and SilverPush are the main providers of the uXDT technology, of which SilverPush is by far the largest.[650] According to German researchers at Technische Universität Braunschweig, there were only six SilverPush uXDT-enabled apps in April 2015. By December 2015, the number increased to thirty-nine. As of May 2017, there were 234.

On its website, SilverPush pitches its invasive technology to prospective clients with unabashed candor. "Next time don't just reach your customers, know them," it declares. "Next generation Advertising 2.0 is here."[651]

SilverPush is tightlipped about its uXDT clients, so we can't be sure which apps are spying on us with ultrasonic beacons—although it appears none are currently operating in the United States.[652] For the aforementioned German researchers, SilverPush's high level of secrecy indicates "the step between spying and legitimately tracking is rather small."[653]

Of course, we can easily protect against uXDT snooping by refusing to give apps permission to use our smartphone's microphone. But app makers know we're not liable to do that. Even if suddenly we all did, app makers would only need to offer deep discounts to anyone willing to play ball, and no doubt many of us would do just that.[654] [See SPY: MEMORY LANE.]

The unavoidable reality is smartphones, boon that they surely are, have become "small tracking devices," says Michelle De Mooy, acting director of the Center for Democracy and Technology's Privacy & Data Project. "We may not think of them like that because they're very personal devices—they travel with us, they sleep next to us. But they are in fact collectors of a vast amount of information including audio information."[655]

## Robot Vacuum Cleaners

Robot vacuum cleaners rely on a system of infrared sensors, laser light, and in some cases low-resolution cameras to clean floors without bumping into things or getting cornered. It's called *simultaneous localization and mapping* (SLAM) technology.

A robot vacuum uses the technology to create digitized maps of a home's interior. That way, if it runs low on power, for instance, it can stop, find its way back to the recharging station, plug itself in until fully powered up, then return to where it was and resume the job.

In 2017 Reuters revealed iRobot—the Massachusetts-based manufacturer of the popular Roomba vacuum cleaners [see ROBOT: MEET YOUR NEW BFF]—plans to share the maps of customers' houses with other vendors. Colin Angle, iRobot's CEO, defended the plan by noting it's the *customers* who are allowing such sharing to occur. "There's an entire ecosystem of things and services that the smart home can deliver, once you have a rich map of the home that the user has allowed to be shared."[656]

The user has allowed.

Check.

Colin Angle counts on customers mindlessly checking off the "I Agree" box next to the EUL for iRobot's HOME app. And why shouldn't he? The app allows us to control Roombas from anywhere in the house with just our smartphones. Why would we deny ourselves that singular convenience, especially after paying up to $800 for the IoT gadget?[657]

Thus, Angle is technically correct. It is entirely with our knowledge and consent that the little spies rove about, vacuuming up everything they can learn about us and our domiciles and storing it in a cloud.[658]

Angle denies iRobot is *selling* the personal information to third parties, piously averring in a written statement, "iRobot will <u>never</u> sell your data."[659] Instead, he says, iRobot is giving the data away for *free*.

But Angle's giveaway scheme is crafty. Here's how it works.

He plans to give away the Roomba maps of customers' homes to the makers of other smart-home devices. They include *smart stereo systems* that conform optimally to a room's acoustics; *smart climate-control systems* that customize airflow to each room, according to its needs; *smart lighting systems* that control for a room's windows, the season, and time of day; and *smart security systems* that know where everyone and everything is in a house at any given time.

In return for Angles' free intelligence, these other vendors will naturally be incentivized to make their smart-home devices uniquely compatible with iRobot's products and the maps of our houses. This inevitably will increase iRobot's share of the market and profits.[660]

### Wireless Headphones

Bose, famous for its high-end stereo systems, sells a suite of enormously popular wireless headphones. They include the QuietComfort, QuietControl, SoundSport, and SoundLink product lines.[661]

Like nearly all IoT devices, the wireless headphones come with a smartphone app—which, yes, features a finely printed EUL and prominent "I Agree" box. The Bose Connect app enables customers to, among other things, easily curate and control their song libraries via their smartphone touchscreens.

But, hold on.

In a federal, class-action lawsuit filed in Illinois's Northern District on March 18, 2017, attorneys allege: "Unbeknownst to its customers . . . Defendant designed Bose Connect to (i) collect and record the titles of the music and audio files its customers choose to play through their Bose wireless products and (ii) transmit such data along with other personal identifiers to third-parties—including a professional data miner—without its customers' knowledge or consent."[662]

On its webpage, Bose insists: "Bose Corporation ('Bose') respects the privacy of our users."[663] But then it immediately lists all the information it collects on users:

- Your software and hardware attributes—including Device operating system version, MAC address, and hardware model information
- Device identifiers such as IP address and Device IDs
- Your time zone
- Information about when you use the App—such as date and time and duration
- Network information such as network type and carrier
- Information about the Bose Products you connect to the app—such as Bose Product model, name, serial number, and product settings (e.g., volume, bass/treble level)
- How you use the App during your current session and over time, including the media content to which you connect while using the App (e.g., song or podcast title, artist, and playing time)

And then, as a postscript, Bose discloses this gem:

"To gather the information discussed in this section, we or our service providers may use web logs or applications that recognize your Device and gather information about your use of the App, including software developer kits ("SDKs"), pixels, scripts, or other tracking mechanisms."

That's what now passes for respecting a users' privacy.

"Indeed, one's personal audio selections—including music, radio broadcast, Podcast, and lecture choices—provide an incredible amount of insight into his or her personality, behavior, political views, and personal identity," reads the lawsuit. "In fact, numerous scientific studies show that musical preferences reflect explicit characteristics such as age, personality, and values . . ."[664]

Segment, a San Francisco-based data-mining firm, is allegedly one of the third parties receiving all that personal information from Bosc's headphone apps. On its website, Segment—which boasts having 15,000-plus clients—claims it can "capture data from every customer touch point" and then send it to wherever "it can be used most effectively."[665]

Put bluntly, Segment collects and processes personal information from a legion of IoT snoops then pimps it out to the world's highest bidders.

*Listening Devices*

The age of IoT lends fresh meaning to the old admonition, "Be careful what you say; the walls have ears." Today, ears are literally everywhere, having become a big part of the smart-technology boon.

Our *cars* have ears, with voice-activated systems like Dragon Drive,[666] Ford Sync,[667] and GM IntelliLink.[668] Our *children's bedrooms* have ears, with popular sound monitors by VTech, Safety 1st, and Motorola.[669] Even *wristwatches* such as Fitbit,[670] Apple Watch,[671] and CoWatch have ears.[672]

Consider, too, the voice-activated assistants in our homes. Amazon Echo,[673] Google Home,[674] and Apple HomePod[675] are smart, modern-day genies that are always all-ears [see ROBOT: MEET YOUR NEW BFF].

But how smart are *we* for buying into it all?

Without her parents' knowledge or permission, a six-year-old Dallas, Texas, girl used an Echo Dot to order herself a KidKraft Sparkle Mansion dollhouse and a large tin of sugar cookies. When they found out, her parents donated the mega-dollhouse to a local children's hospital.[676] Now, the entire family watches what they say around the big-eared Echo Dot. "I [now] like whispering in the kitchen," confesses the girl's mom.[677]

Remarkably, the saga doesn't end there.

The story aired on the CW6 News' morning show in San Diego, ending with anchor Jim Patterson quipping, "I love the little girl saying, '*Alexa* ordered me a dollhouse.'" Immediately, Echo devices all over the city hearkening to the anchor's voice each began attempting to do the very same thing—order a dollhouse.[678]

You see, the technology is relentless and unforgiving. The minute we say the magic *wake-word*—in this case, "Alexa,. . ."—the smart device instantly fires off our verbal request via the web to a cloud, where our words are analyzed, and the genie's response is formulated. In effect, the cloud is the genie's brain.

Typically, an audio file of all our requests is permanently stored in the cloud. It helps the genie remember our preferences, as well as the idiosyncrasies of our voice.

There are ways to disable this recording feature to protect our privacy. But, here again, few of us do that, because it kneecaps the genie. Whereupon it becomes, as one *Wired* reporter wryly put it, little more than a paperweight.[679]

*Kids Toys*

In July 2017 the FBI issued a stern warning to parents about smart toys. "These toys typically contain sensors, microphones, cameras, data storage components, and other multimedia capabilities—including speech recognition and GPS options," it said. "These features could put the privacy and safety of children at risk because of the large amount of personal information that may be unwittingly disclosed."[680]

That's precisely what happened with the web-connected stuffed animals called CloudPets, made by a California company named Spiral Toys. Briefly, here's how the plush toys work:

1.  Using the CloudPets' smartphone app, a loved one can send messages to a child from anywhere in the world.
2.  A grownup at home receives each message, approves it, and forwards it wirelessly to the child's CloudPet.

3.  The CloudPet's heart responds by blinking, whereupon the child squeezes the animal's paw to hear the message.
4.  The child can record a response by again squeezing the paw. The message is then transmitted directly via a nearby smart device to the loved one anywhere in the world.[681]

In early 2017 several security experts—including Australian Troy Hunt, a Microsoft Regional Manager—gained access to an unsecured cloud server containing the account information of more than 800,000 Cloud-Pets users. The trove of stored data included children's names, photos, email addresses, and 2.2 million recorded personal messages.[682]

Parents "probably didn't think through the fact that when you connect the teddy bear [to the web]," says Hunt, "your kids' voices are sitting on an Amazon server."[683] It was information anyone could find simply by using the IoT search engine Shodan.[684]

Spiral Toys maintains no personal information was stolen and, in any case, has since then fixed the problem. "To our best knowledge, we cannot detect any breach on our message and image data, as all data leaked was password encrypted."[685]

But Hunt and the other experts disagree, citing evidence the database was stolen more than once by hackers and held for ransom. They also point out many of the passwords were extremely weak and easily guessed. "Anyone with the data could crack a large number of passwords, log on to accounts and pull down the voice recordings," explains Hunt.[686]

Red flags also have been raised concerning IoT dolls Hello Barbie and Cayla, as well as hereO kids' watches.[687] The list of hackable toys is quickly growing, says the FBI, thanks to companies chasing after a quick buck. "Security safeguards for these toys," it warns, "can be overlooked in the rush to market them and to make them easy to use."[688]

## Adult Toys

Couples using We-Vibe vibrators—made by Ottawa-based Standard Innovation—can control one another's smart dildos via the web. "Turn on your lover when you connect," the product website says, "and play together from anywhere in the world."[689]

The We-Connect smartphone app lets you control your partner's vibrator, with rousing modes like the pulse, wave, crest, peak, and cha cha cha. Also, says the website, the app lets you "Build excitement with secure in-app voice, chat and video."[690]

But there's a problem with all of it.

In late 2016 a pair of hackers, codenamed g0ldfisk and followr, discovered roguish voyeurs standing within thirty-five feet of the smart dildos—say, just outside the house—could hijack them via Bluetooth and make them dance to their tune, so to speak.[691]

And that's not all.

An Illinois woman identified simply as "NP" filed a federal class-action lawsuit against Standard Innovation, claiming: "Defendant programmed We-Connect to secretly collect intimate details about its customers' use of the We-Vibe, including the date and time of each use, the vibration intensity level selected by the user, the vibration mode or pattern selected by the user . . . and incredibly, the email address of We-Vibe customers who had registered with the App, allowing Defendant to link the usage information to specific customer accounts."[692]

"It's one matter collecting data about your usage of a smart coffee machine," remarks Ken Munro of Pentest Partners, a British firm that tests the security of IoT devices. "It's a whole different matter gathering data about your sex toys."[693]

In a settlement announced in March 2017, Standard Innovation agreed to pay NP and the other plaintiffs a total of $5 million in Canadian currency—about $3.9 million in American money.

The company also agreed to "(i) purge email addresses provided by application users as part of the We-Connect application registration process . . . and (ii) purge the following data elements: the time and date of each use, the vibration intensity level selected by the user, the vibration mode or pattern selected by the user, the temperature of the device, and the battery life."[694]

It was a fitting climax, one might say, to a very heated matter.

## THE INTERNET OF EVERYTHING

At this point, you might be thinking: Who needs secret police, when so many IoT devices are spying and snitching on us? And doing so lawfully. With our blessing.

"You know something is wrong when the government declares opening someone else's mail is a felony," quips author and blogger Elizabeth Ann Bucchianeri, "but your internet activity is fair game for data collecting."[695]

If you're wondering how much information about your internet behavior is being noticed, here's an easy experiment you can do.

If you have a Google account—and who doesn't these days?—log into https://myactivity.google.com/ and behold all the information that is stored about your online activity. It includes topics you've searched, websites you've visited, and videos you've watched.

And that's just Google, which is relatively transparent about the lowdown they have on us. Think of all the IoT gadgets with which we interact daily, each one stashing away juicy tidbits of highly personal information about us for other people to see—total strangers halfway around the world, even.

For children now being born, their electronic dossiers begin right away. Hospitals digitize medical records and exchange them online. And schools expect kids to use electronic tablets for notetaking and the web for research.

Gennie Gebhart, a researcher at the San Francisco-based Electronic Frontier Foundation, warns that today's surveillance culture "threatens to normalize the next generation to a digital world in which users hand over data without question in return for free services, a world that is less private not just by default, but by design."[696]

It's a world run by what experts are calling the *Internet of Everything*, or IoE. A world that jeopardizes not just our personal privacy, but our civilization's very existence.

On October 21, 2016, we got an unsettling peek at this dubious IoE world, when web service came to a screeching halt throughout North

America and Europe. Twitter, Spotify, Reddit, *The New York Times*, Pinterest, PayPal, and other major websites went completely dark.

The culprit? Mirai, a nasty bit of malware that infects IoT devices. On that eye-opening October day, hundreds of thousands of Mirai-infected security cameras, routers, and DVRs helped spread their contagious disease throughout the world wide web at the speed of light.[697]

In 2017, in a report ominously headlined, "A New IoT Botnet Storm is Coming," Check Point—a major Northern California cybersecurity company—issued a warning about a vicious new malware named IOTroop. The report's three subheads read:

- A massive Botnet [IOTroop] is forming to create a cyber-storm that could take down the internet.
- An estimated million organizations have already been infected.
- The Botnet is recruiting IoT devices such as IP Wireless Cameras to carry out the attack.

According to the report, IOTroop is "evolving and recruiting IoT devices at a far greater pace and with more potential damage than the Mirai botnet of 2016." It appears to be exploiting "vulnerabilities in Wireless IP Camera devices such as GoAhead, D-Link, TP-Link, AVTECH, NETGEAR, MikroTik, Linksys, Synology and others."[698]

Welcome to the beginnings of the IoE world.

Way back in 2008, a cheery IBM predicted the IoE would usher in the era of a "Smarter Planet," comprising everything "from smarter power grids, to smarter food systems, smarter water, smarter healthcare and smarter traffic systems." The IoE would surely create a tsunami of data, said IBM, but not to worry, the company was already busy creating algorithms to make sense of it all.[699]

Today's nascent IoE—with its flood of information, data, analytics, and algorithms—is, of course, great news for IBM's bottom line. It spells unlimited j-o-b s-e-c-u-r-i-t-y for the company's super computers, such as Watson. But for the rest of us, *big data* and *big algorithms* foreshadow *big brother.*

Think about it: smart devices can only become "smarter," as IBM foresaw, by learning more about us. Not just our voices, faces, behaviors, preferences, politics, and sex lives, but *everything.* For that reason, I honestly can't see how the idea of an IoE—of a "smarter planet"—is ultimately good for *our* bottom line.

"The internet, our greatest tool of emancipation, has been transformed into the most dangerous facilitator of totalitarianism we have ever seen," cautions Julian Assange, founder of Wikileaks. "Left to its own trajectory, within a few years, global civilization will be a postmodern surveillance dystopia . . . In fact, we may already be there."[700]

# OPERATING IN THE
# SHADOWS

*"In the kingdom of glass everything is transparent,*
*and there is no place to hide a dark heart."*

Vera Nazarian

Privacy was not always considered a good thing. During the Middle Ages, explains Alice Jane Cooley in her PhD thesis at the University of Toronto's Centre for Medieval Studies, "there was a heightened suspicion of privacy as a state conducive to illicit activity and facilitative of sin."[701]

Today, centuries later, the concern is still relevant. Consider the many crimes still being committed in secrecy: murder, robbery, vandalism, blackmail, embezzlement, you name it. In such cases, privacy-busting measures—from x-raying airline passengers to microchipping violent criminals—are needed to help maintain law and order.

Sherlock Holmes, fictional literature's most famous sleuth, practiced his craft between 1880 and 1914, when the concept of a crime scene was just becoming popular. "Holmes throughout his adventures would constantly make reference to how crime scenes were easily contaminated and he always emphasized how maintaining the integrity of the scene was of the utmost importance," notes real-life forensic scientist Robert Ing.[702]

In 1918 French scientist Edmond Locard formulated the rules informing crime scene investigations to this day. His main idea, known as Locard's Exchange Principle, states that in the process of committing crimes, perpetrators inadvertently and inevitably leave behind and/or take with them bits of incriminating evidence. If gathered and analyzed properly, that body of evidence can be used to track a criminal down.[703]

As the crime scene's importance has grown, forensic technology has greatly improved. Real-life detectives in Holmes's day used magnifying glasses to analyze hair and fabric samples, microscopes to distinguish bloodstains from generic stains, and chemical tests to differentiate human blood from animal blood, as well as to identify a wide variety of poisons within body tissue. Those early detectives were also among the first to bring criminals to justice by using fingerprints, blood types, and the telltale scratches fired guns create on bullets.[704]

Today, more than ever, science is the enemy of a criminal's privacy and hopes to evade arrest and prosecution. As attorney Nicole Chauriye puts it, "Technology is killing our opportunities to lie."[705]

Here are some of today's privacy-busting technologies helping to out even the sneakiest behavior.

*Digital Cameras*

- **Eye in the Sky.** In 2016 a man using the screen name YAOG claimed he was tipped off by a friend that his wife of eighteen years was cheating on him. Twice, he tried tailing her on foot, as she walked to work. But to no avail. That's when he had the idea of using a tiny video camera aboard a high-flying drone to get at the truth. Sure enough, he says, it worked: the aerial spy caught his wife appearing to rendez-vous with, kiss, and get into the car of a local Romeo.

YAOG posted the video of his allegedly unfaithful wife on YouTube for all the world to see. "Here she is waiting at the intersection, waiting at the lights for what seems like forever, and here she is taking her hair down," he says, in a blow-by-blow account. "If you watch carefully, you'll be able to see eighteen years [of marriage] go down the f---ing drain!" To date the video has racked up more than *fourteen million* views.[706]

In follow-up YouTube postings, YAOG claimed he and his wife were legally separating, despite their having young children and her purportedly being contrite and wishing for a reconciliation. "I told her that can't happen," he reports in a YouTube video titled, Marriage Update #2.[707]

In November 2017, however, the couple—John and Donna from Honesdale, Pennsylvania—went public, appearing on the TV show *Inside Edition* to say they were reconciling. Ironically, John credits the spy-in-the-sky for saving his marriage. "If I hadn't sent that drone up and saw what happened," he says, "I believe that the situation [Donna's budding extramarital affair] would have gotten more intimate." He adds, "People aren't perfect. People deserve second chances."[708]

- **Eye in the Classroom.** In late 2016 Caleb O'Neil, a nineteen-year-old freshman at Orange Coast College in Southern California, was sitting in his human sexuality class, listening attentively to Professor Olga Perez Stable Cox. Without warning, she launched into a political tirade about the election of President Donald Trump. "We have been assaulted," Cox told her young captive audience. "It's an act of terrorism."[709] She reportedly finished up by asking Trump supporters in the classroom to stand up, then called them "cowards" when they did not.[710]

Using his smartphone, O'Neil videotaped most of the rant. "I pulled my phone out," he recounts, "because I was honestly scared that I would have repercussions with my grades because she knew I was a Trump supporter."[711]

O'Neil showed the videotape to the school's chapter of College Republicans and to administration officials. Unhappy with the administration's slow reaction, Joshua Recalde-Martinez of the College Republicans posted Cox's political diatribe on YouTube, where it went viral. "At this point it's not even education anymore," says Recalde-Martinez about Cox's polemic, "its indoctrination."[712]

At first, Orange Coast College responded by supporting the professor and suspending O'Neil, claiming he had no right to videotape Cox without her consent. But after facing a firestorm of criticism by O'Neil's attorneys and the public, the college reversed its decision and reinstated the freshman.[713]

- **Eye on the Toilet.** In 1420, on 667 acres of what is now central Beijing, Chinese Emperor Yongle built a unique campus of imperial religious buildings. The now-popular, Ming Dynasty tourist attraction is called the Temple of Heaven Park.[714]

  In 2007 the park began offering free toilet paper in its public restrooms. But lately, too many people—especially the elderly, it is claimed—have been stealing the toilet tissue in large quantities.[715]

  In 2017 the communist government put its foot down and installed high-definition, facial-recognition cameras to thwart the bad actors. Bathroom users must now look into the camera for three seconds, after which the smart device dispenses exactly two feet of toilet paper.

  Need more?

Wait nine minutes, then look again into the camera, and you'll get another two feet of tissue. Unless the camera's AI software nails you as a repeat offender, in which case, you get nothing.[716]

After the first two weeks of operation—during which, predictably, there were many glitches—park officials reported toilet paper consumption plunged by 20 percent. The savings prompted the government to offer visitors two-ply tissue, instead of the standard-issue, single-ply paper.[717]

For many, however, the small luxury does not nearly offset the ignominy of this latest high-tech intrusion into their personal lives. "I thought the toilet was the last place I had a right to privacy," lamented one park visitor, "but they are watching me in there too."[718]

## DNA Phenotyping

In 2009 Sheriff Tony Mancuso of Louisiana's Calcasieu Parish had only two clues to the brutal murder of nineteen-year-old Sierra Bouzigard: tissue found under her fingernails, presumably from the assailant; and the number of her last phone call.

The number led Mancuso to a band of immigrant Mexican laborers, whose DNA he collected. But none of the samples matched the DNA from under the victim's nails. Neither did Mancuso score a bullseye with the more than sixteen million DNA profiles the FBI keeps in its Combined DNA Index System (CODIS).[719]

As a result, the case went cold.

Six years later the sheriff reached out to Parabon NanoLabs, a company in Reston, Virginia, which claimed it could generate photo-likenesses of suspects based solely on their DNA. Parabon's novel technique, developed with the support of the US Department of Defense, is based on some straightforward biology.[720]

Our cells contain strands of DNA whose 3.2 billion molecules control the basic functions of the mind and body common to all of us. But in at least *ten million* places along the DNA strands, there are slight variations. They are what account for our individual differences.[721]

With its proprietary Snapshot DNA Phenotyping System, Parabon is able to identify tens of thousands of these telltale variations in a person's DNA—from which it deduces key identifiers.[722] These include, says Parabon's website, a person's "ancestry and physical appearance traits, such as skin color, hair color, eye color, freckling, and even face morphology."[723]

Sheriff Mancuso was shocked when he received Parabon's analysis of the DNA collected from under Bouzigard's fingernails. According to it, the presumed perpetrator was *not* from Mexico at all, but most likely from Northern Europe, and probably had pale skin, freckles, brown hair, and either green or blue eyes. "We kind of had to take a step back and say all this time, we're not even in the right direction," remarked the sheriff.[724]

In July 2017, the new information led Mancuso to Blake A. Russell, a 31-year-old Caucasian man who looked remarkably like Parabon's computer-generated snapshot. (See for yourself, by checking out Parabon's website: https://snapshot.parabon-nanolabs.com/posters.)[725] Russell's DNA matched the DNA from under Bouzigard's nails, leading Sheriff Mancuso to charge Russell with second-degree murder.[726]

## Virtual Wiretapping

Under normal circumstances, police authorities cannot eavesdrop on our private conversations without approval from a judge. But with the proliferation of voice-activated assistants, we are effectively wiretapping ourselves [see ROBOT: MEET YOUR NEW BFF and SPY: OMNISCIENT OBJECTS OF OUR DESIRE]

Twenty-eight-year-old Eduardo Barros found it out the hard way, in the summer of 2017. While he and his girlfriend were taking care of her parents' home in Tijeras, New Mexico, they got into a fight. Barros allegedly accused her of cheating on him and, whipping out a gun, warned if she called the police, he'd kill her right then and there.[727]

During the escalating fracas, his suspicions about her boiled over and he demanded to know, "Did you call the sheriffs?" Unbeknownst to him, a voice-activated assistant—the police report is unclear about which kind, but apparently it was somehow patched into the house's surround-sound stereo system—mistook the words "call the sheriffs" as a command to do just that.[728]

SWAT and police negotiating teams quickly converged on the house and, hours later, took Barros into custody, charging him with fourteen crimes, including aggravated battery against a household member and possession of a firearm by a felon. "This amazing technology," marvels Bernalillo County Sheriff Manuel Gonzales III, "definitely helped save a mother and her child from a very violent situation."[729]

### Virtual Stakeout

On December 23, 2017, firefighters responding to a 911 call rushed to a house in Ellington, Connecticut, where they found Richard Dabate lying on the kitchen floor, tied to a chair and bleeding. Minutes later they found his wife in the basement, dead from a gunshot wound.

For hours the police questioned Dabate about what happened. Here's what he told them.[730]

During his drive to work that morning, his cell phone indicated the house alarm had gone off. Immediately, Dabate turned the car around and sped home, firing off an email to his boss saying he'd be late for work.

When he arrived at the house, Dabate dashed up to the master bedroom, where he was confronted by a six-foot two-inch masked vandal with a voice like Vin Diesel's.[731]

Moments later Dabate heard his wife, Connie, returning from exercise class. He yelled at her to run, which she did. But the intruder chased her downstairs, with Dabate following close behind. The intruder caught up with Connie in the basement, shot her in the back of the head, then overpowered Richard, zip-tying him to a folding chair before fleeing.

Nearly unconscious, Dabate managed to drag himself and the chair upstairs to the kitchen. There, he triggered the house's panic alarm and called 911, before passing out.

The Dabates' neighbors were shocked to hear the news, all of them telling police Richard and Connie were a loving couple, the very picture of marital bliss. "They couldn't keep their eyes off each other," one neighbor reported. "It was a look that you would want."[732]

In an attempt to reconstruct the events of that tragic day, police collected information from a number of unlikely sources—including Connie's Fitbit, which they found on her body. Fitbits—which have become extremely popular in recent years—are wearable smart devices capable of monitoring people's activity, exercise, food consumption, weight, sleep, and much more.[733]

According to the stored data on Connie's Fitbit, she walked 1,217 feet after driving home from exercise class. This immediately caught the authorities' attention, because Richard's version of events had her fleeing only about 125 feet, the distance from garage to basement. Moreover, the Fitbit indicated Connie was still active nearly an hour *after* she was supposedly killed. Her Facebook page, too, showed her actively posting videos about forty-five minutes *after* she was purportedly lying dead in the basement.

From there, things went from bad to worse for Richard Dabate.

Data stored by the house's security system showed the alarm that went off that morning was triggered from the basement by Richard's key fob. And Richard's Microsoft Outlook account revealed his email to the boss—allegedly sent from Richard's car—was actually sent from an IP address corresponding to a computer inside the house.

Based on the strength of this incriminating evidence, police arrested Richard Dabate on charges of murder, tampering with evidence, and making false statements.[734] He has pled not guilty and as of this writing is free on $1 million bail.[735]

## CLOSING THOUGHTS

"All human beings have three lives: public, private, and secret."[736] Alas, that seemingly tautological observation, made by Nobel Prize-winning Colombian writer Gabriel García Márquez, is no longer true. Today, the traditional walls between public, private, and secret are all but gone.

What's more, there is an exact moment in history when the walls suffered a massive blow and began tumbling down. That moment was September 11, 2001.

Immediately after the 9/11 terrorist attacks, the stocks of most American companies tanked. But those of Visionics Corporation, a Minnesota-based company, skyrocketed. Why? Because their facial-recognition program, FaceIt, suddenly became just what the doctor ordered.[737]

"Recent events," declared Visionics CEO Joseph J. Atick on October 3, 2001, "have demonstrated that an intelligence-based identification system is needed to identify criminals and terrorists. A key component for building this shield is a system like FaceIt."[738]

"Before Sept. 11," observes Jeffrey Rosen, CEO of the National Constitution Center, "the idea that Americans would voluntarily agree to live their lives under the gaze of a network of biometric surveillance cameras, peering at them in government buildings, shopping malls, subways and stadiums, would have seemed unthinkable, a dystopian fantasy of a society that had surrendered privacy and anonymity."[739]

Yet, that's precisely what happened.

Today, as we've seen in the chapters of this section, we live in a world surveilled by cameras, social media, and an army of electronic snoops, many of which comprise the so-called Internet of Things—soon to become the Internet of Everything. These web-connected smart devices quietly eavesdrop on our everyday lives and are mostly under the command of total strangers: corporations, governments, and hackers.

But also, our next-door neighbors.

These smart devices—too clever by half—watch us, listen to us, and evaluate our behavior constantly. And they do so, for the most part, *with our permission*. All in the name of entertainment, convenience, and, yes, *security*—as today's post-9/11 American surveillance state vividly illustrates.

Still, the fight for what remains of our privacy is far from over.

Case in point: In 2016 the Fresno police department created a stir when dispatchers in its science-fiction–like Real Time Crime Center began using Beware, a new AI computer program, made by West Corporation in Omaha, Nebraska. Among other things, Beware tags each of us with a color-coded threat level—green, yellow, or red—based on billions of records housed on the web, including our criminal histories, website visits, and social media postings.

During the program's trial period in Fresno, police dispatchers fed Beware's color-coded threat assessments and other personal intel to officers responding to 911 calls.[740]

"Beware is location-based and quickly provides names of individuals who are most likely residing at the addresses where officers are responding," explains Fresno police chief Jerry Dyer. "In addition, Beware can provide contact phone numbers, relatives and their contact numbers, criminal history, and hyperlinks to any public social media posts found on the Internet connected to an individual."[741]

Not everyone was impressed with Beware's prowess.

"At what point," says the ACLU's Jay Stanley, "does that begin to resemble China's incipient 'citizen scoring' system, which threatens to draw on social media postings and include 'political compliance' in its credit-score-like measurements?"[742]

The Fresno public's opposition to Beware was so vociferous, the city council immediately pulled the plug on it. Nonetheless, as of this writing, Beware and its manufacturer are still very much in business.[743] So, beware.

As I wrap up this discussion, I see no end in sight to technology's unprecedented assault on our privacy. Back in 1966, Supreme Court justice William O. Douglas foresaw "the fantastic advances in the field of electronic communication constitute a great danger to the privacy of the individual."[744]

His prophecy has come true.

In spades.

At this very minute, for instance, psychologist Marcel Just and his colleagues at Carnegie Mellon are marrying brain-imaging technology with AI to read our *minds*.[745] They are doing so by having people speak a sentence and noting the brain pattern it causes.

One day we'll be able to use the information in reverse: by looking at someone's brain patterns, we'll be able to know what the person is thinking.[746] In other words, Just's research is gradually producing an Orwellian dictionary, which will let us look up the meanings of human brain patterns.

To be sure, the motive of the Carnegie Mellon researchers is high-minded. They wish to read the thoughts of suicidal patients, in hopes of intervening in time to save lives.[747]

But as we've seen time and again with science's noblest intentions, there is nearly always the devil to pay on account of unintended consequences. Here, for example, it doesn't take much imagination to foresee how corporations and governments will be able to exploit Just's Orwellian dictionary for their less-than-noble purposes.

"I can't imagine a system [designed] to take value readings of my mind for a remote company being used for good," says IT specialist Chris Dancy about mind-reading technology in general. "It's a dark path."[748]

Despite its good intentions, Carnegie Mellon's research clearly represents technology's vanquishing blow to privacy's last stand. For, as Pulitzer Prize-winning author Annie Dillard observes in *The Maytrees*: "Where is privacy, if not in the mind?"[749]

Where, indeed?

# FRANKENSTEIN

*"Men ought not to play God before they learn
to be men, and after they learn to be men
they will not play God"*

Paul Ramsey

# MEMORY LANE

*"Renegade scientists and totalitarian loonies
are not the folks most likely to abuse genetic
engineering. You and I are—not because we are
bad, but because we want to do good."*

Arthur Caplan

n the beginning, so the Bible reports, God gave us the most precious
gift of all: free will. Adam and Eve cashed in on that freedom and the
rest, as they say, is history.

Nowhere is our storied willfulness more evident than in our relationship with nature, God's creation. Over the millennia, our choices—our
myriad preferences for certain plant and animal traits—have systematically recreated the living world in our own image.

Some call it playing God. Others see it as we humans simply exercis-
ing, for better or worse, our God-given free will.

As I'll explain in this section, both claims are true.

But what's also true—and is often overlooked in today's heated
debates about genetic engineering—is nature is not static. Natural forces,
and we along with them, have been constantly meddling with the original world long before science was ever born.

## NATURE'S LIFE-AND-DEATH DECISIONS

Long before the theory of evolution, our ancient ancestors could plainly see some plants, animals, and people are hardier than others. They're better able to survive adversities—harsh winters, famines, droughts, storms, diseases, climate change—and can even turn them into opportunities for advancement.

That, in a nutshell, is Charles Darwin's concept of natural selection. A concept he spelled out in his famous tome *On the Origin of Species by Means of Natural Selection, or the Preservation of Favoured Races in the Struggle for Life*, published in 1859.

In the face of an ever-changing environment, Darwin argued, a single trait can "determine which individual shall live and which shall die—which variety or species shall increase in number, and which shall decrease, or finally become extinct."[750]

Darwin claimed something else: nature is constantly and randomly introducing *new traits* we pass on to our offspring. We now know these variations are caused, for example, by glitches in the normal reproduction process and by cosmic rays, toxic chemicals, and other environmental, mutagenic forces constantly dinging our DNA.[751]

These natural sources of new traits perpetuate a diversity among living organisms. They increase the odds at least some individuals of a species will have the "right stuff" to fend off whatever nature throws at them.

The fabled American Chestnut tree, for instance, once carpeted the eastern United States from Maine to Florida, and from the Piedmont plateau to the Ohio Valley. During the first half of the twentieth century, nearly all the estimated four billion trees on 200 million acres were decimated by a fungus accidently imported from Japan.[752]

The loss was devastating, because the tree's unusually hard, rot-resistant wood was widely used to build log cabins, furniture, fence posts, railroad ties, you name it. Also, turkeys, bears, raccoon, deer, and squir-

rels relished the chestnut tree's sweet, flavorful nuts—as did we, especially roasted at Christmastime.[753]

"I think of it as the miracle tree," says Tom Klak, a professor at the
University of New England, in Saco, Maine. He's one of many people
now working hard to reverse the American Chestnut's tragic fortunes.[754]

A reversal is possible in part because a few American Chestnut trees
chanced to have traits that protected them from the devastating blight.
Several years ago one such lucky survivor was spotted from the air, growing in a forest in Lovell, Maine, a tiny town of 1,140 residents.[755]

In December 2015, Brian Roth, a University of Maine arborist,
trudged through the forest on foot and helped confirm the extraordinary find. "We think it's around one hundred years old," he says. He
and his colleagues also determined that the hearty survivor is 115 feet
high, making it the tallest known American Chestnut living in
North America.[756]

By propagating this tree's blight resistant traits—along with those
of other known survivors—scientists nationwide hope to do the seemingly impossible. "We see real promise of the future to bring the species
back," says Lisa Thompson, president of the American Chestnut
Foundation.[757]

## OUR LIFE-AND-DEATH DECISIONS

Some people believe natural selection created living organisms from
scratch. Others believe *God* created the major templates of life, the
basic species, which the process of natural selection now constantly
tweaks.

Either way, surely we can all agree natural selection is an elegant way
to maintain the health and wellbeing of living organisms struggling to
survive on a ferociously dynamic planet. The alternative—a completely
*static* world, where lifeforms never change—is not realistic, healthy, or
at all interesting.

"There is no permanent status quo in nature; all is in the process of adjustment and readjustment," observed the late Nobel Prize-winning American geneticist Hermann Joseph Muller. "But man . . . has the power to note this changefulness, and, if he will, to turn it to his own advantage."[758] For better and worse, that is precisely what we have done, for as long as our species has existed.

Look around you. Neither natural selection nor God created many of the living organisms we most cherish. *We* did.

Before we got our hands on them, for instance, "carrots were white and spindly. Wheat was tall and scrawny with little calorific value," explains James Kennedy, a celebrated science instructor at Australia's Haileybury Institute. "Apples were tiny and sour with giant pips (like crab-apples today). Strawberries were tiny, bananas had stones in them, and pigs were vicious creatures with tiny backsides that made for a not-so-delicious ham. Cows didn't produce much milk (just enough for their own calves), and chickens were skinny little creatures that laid eggs weekly rather than daily."

Left up to natural selection or the Almighty, tomatoes would still be the size of peas, corn would be little more than a grass head, and dogs wouldn't exist. These organisms—and many, many others—are *our* creations. The result of thousands of years of *artificial selection*—of unconscious and deliberate human choices.

Here are some historic examples of how, by trumping natural selection and God, artificial selection systematically produced three of our all-time favorite things.

### Tomatoes

In prehistoric times, wild tomatoes—no bigger than shelled peas—grew aplenty in the Andes mountains of Peru, Ecuador, and Bolivia. About two thousand years ago, natives carried the straggly vines northward to Mexico.[759] Then, in the sixteenth century, Old World explorers transported them across the Atlantic Ocean to Europe.[760]

Italians were among the first to develop a taste for the tiny red fruits—along with the French, who called them *pommes d'amour*, or "love apples." Early on, the British grew tomato plants only as ornamentals, believing the fruits to be poisonous. After all, they do belong to the nightshade family.[761]

For two hundred years, those early Europeans artificially selected and promulgated traits of the wild tomato that, among other things, greatly swelled its size. Then, during the eighteenth and nineteenth centuries, when hordes of Europeans migrated to the United States, the new varieties came with them.

We can only imagine how astonished native Americans were when they first laid eyes on the fruit. Their beloved lowly, pea-sized tomato had returned to the New World—*one hundred times larger!*[762]

### Corn

Nine thousand years ago, southern Mexico's Balsas River Valley was blanketed by a wild grass called teosinte. According to most scientists and historians, native Americans living there used artificial selection to steadily transform the plentiful grass into maize, AKA corn.[763]

"The most impressive aspect of the maize story is what it tells us about the capabilities of agriculturalists 9,000 years ago," remarks Sean Carroll, a biologist at the University of Wisconsin–Madison. "These people were living in small groups and shifting their settlements seasonally. Yet they were able to transform a grass with many inconvenient, unwanted features into a high-yielding, easily harvested food crop."[764]

Compared with those of the lowly teosinte, the kernels of today's corn are (1) one-thousand times larger, (2) three times sweeter, and (3) many times softer. Moreover, corn is now the most prolific grain crop in the world,[765] comprising one-fifth of the global human diet.[766]

"It is grown successfully in every continent but Antarctica," note Ken Russell and Leah Sandall, agronomists at the University of Nebraska-

Lincoln, "from equatorial lowlands to the Matanuska Valley in southern Alaska to Andean highlands that are 12,000 feet above sea level."[767]

*Dogs*

Of all the masterpieces we've created with artificial selection, dogs are one of the oldest and most beloved. Paleontological evidence indicates we bred dogs a very long time ago from a single, docile wolf species—the exact one being a matter of lively debate.[768]

In 2011 an international team of scientists reported discovering the 33,000-year-old "well-preserved remains of a dog-like canid" inside a Southern Siberian cave. After careful analysis, they concluded it is "most similar to fully domesticated dogs from Greenland . . ."[769]

According to the *Fédération Cynologique Internationale* (World Canine Organization), there are now more than 340 different breeds of dog.[770] That means "Man's best friend" gets the blue ribbon for the most diversified mammal on the planet.[771]

## HEART OF THE MATTER

Despite Darwin's historic contribution to biology, he and his generation knew virtually nothing about the actual science of heritable traits, the mechanism that controls the size and shape of, say, a finch's beak or an elephant's trunk. In fact, the subject confounded Darwin greatly.

Darwin believed heritable traits behaved like paint. That if a black cat mates with a white cat, their offspring will be various shades of gray, mixtures of white and black. Yet he knew, in fact, such marriages often produce kittens of pure black or pure white.[772]

The vexing mystery was solved during Darwin's lifetime by an obscure Austrian Augustinian monk named Gregor Johann Mendel, but Darwin and his contemporaries mostly didn't notice and didn't care.[773] In fact, in what surely rates as one of the great missed opportunities in the history of science, even Mendel lost interest in the subject and became the abbot of his beloved St. Thomas's Abbey, in Brno, Czech Republic.[774]

Father Gregor Mendel's simple yet powerful experiments with some 10,000 pea plants are now universally recognized as the founding of modern genetics, which is today making headlines with its breathtaking developments in genetic engineering. "Mendel is a giant in the history of genetics," says David Fankhauser, a biologist at the University of Cincinnati. "I especially admire that he used very simple research techniques that anybody could have duplicated."[775]

From 1856 to 1863, Mendel crossbred many different varieties of pea plants: tall with short; smooth-seeded with wrinkle-seeded; yellow pea with green-pea; and so forth. At the end of each growing season, he carefully noted the traits of the various offspring. By some estimates, in eight years he meticulously tabulated the traits of some 40,000 blossoms and 300,000 peas.[776]

Mendel's grand conclusion?

Traits do *not* mix like paint. Rather, they are determined by heritable particles (we now call them *genes*), each of which comprises a pair of smaller particles (*alleles*) that come in two strengths: dominant and recessive.[777]

Human hairlines, for instance, appear to be controlled by a single gene with two well-known alleles. The allele for a widow's peak (**W**) is dominant; the allele for a straight hairline (s) is recessive. If a fetus inherits the **W** from, say, Dad and the s from Mom—thus inheriting the combination **Ws**—the two alleles will duke it out and the *dominant one will win*. The child will grow up to have a widow's peak.[778]

Altogether, there are four possible combinations of hairline alleles—**WW, ss, Ws, sW**—three of which are dominated by a **W**. Over the long haul, therefore, three out of four fetuses will be born to have a widow's peak. The only times straight hairlines win out are when there is no dominant **W** in the gene—namely, **ss**.

This is the simple yet powerful mathematical reasoning Father Gregor Mendel used to compute the exact proportions of different heritable traits. For that reason, today's biologists call them *Mendelian ratios*.[779]

## HIDDEN SURPRISE

In 1869, just six years after the end of Mendel's unsung experiments, another obscure researcher—Swiss chemist Friedrich Miescher—discovered a strange, new chemical substance dwelling within the nuclei of human white blood cells. He called it *nuclein*.

It was a revelation comparable to Christopher Columbus discovering the New World; yet—shades of Mendel—no one paid much attention. During the first decades of the twentieth century, however, a parade of scientists—among them, Phoebus Levene, Oswald Avery, and Erwin Chargaff—picked up the ball and ran with it. They determined the chemical composition of Miescher's nuclein to be a complex, organic substance called deoxyribonucleic acid (DNA). Moreover, they fingered it as nothing less than the sum and substance of Mendel's heritable particles—genes, alleles, and all.[780]

Curious to know more about it, subsequent scientists pounced on DNA with their cameras, like paparazzi chasing after a Hollywood starlet. By the early 1950s, English researchers Rosalind Franklin, Maurice Wilkins, and Raymond Gosling successfully produced some of the most revealing x-ray photos of the elusive celebrity molecule.[781]

Then on February 28, 1953, came the stunning announcement by American James Watson and Englishman Francis Crick. The molecular biologists successfully puzzled out the structure of DNA and thereby, in Crick's words, "found the secret of life."[782]

DNA, they discovered, is a long, twisting molecule composed of chemical pairs. Picture a long parade of school kids walking two-by-two, each pair holding hands.

Whereas there are two kinds of kids—boy (B) and girl (G)—there are four kinds of basic DNA chemicals: adenine (A), thymine (T), guanine (G), and cytosine (C). Moreover, whereas you can pair kids any way you wish—BG, BB, or GG—with DNA, A always pairs with T, and G always pairs with C. So, a stretch of DNA might look something like AT-AT-GC-AT-GC-GC-GC, and so forth.

Armed with that sensational knowledge, molecular biologists were naturally eager to start mapping the DNA molecules of every known plant and animal species. In 1990, when they finally had the right tools for the job, they set about to diagram the most consequential DNA molecule of all: *the human genome.*

The monumental undertaking—called the Human Genome Project (HGP)—proved every bit as challenging as landing an astronaut on the moon and took thirteen years to complete. But what an historic accomplishment it was, when in April 2003, HGP scientists actually published the human genome's sequence of roughly *3.2 billion DNA chemical pairs.*[783]

"Without a doubt, this is the most important, most wondrous map ever produced by humankind," declared President Bill Clinton. "Today, we are learning the language in which God created life. We are gaining ever more awe for the complexity, the beauty, and the wonder of God's most divine and sacred gift."[784]

## FROM MATCHMAKERS TO GENETIC ENGINEERS

For thousands of years—ignorant of what controlled heritable traits—we were merely matchmakers, breeding A with B, in hopes of getting C. But with the successful completion of the Human Genome Project, we instantly became *genetic engineers*, theoretically able to manipulate DNA molecules directly, precisely, and at will. All we lacked were cheap and effective tools for doing it.

That changed several years ago, when genetic engineers from Harvard, MIT, and UC Berkeley co-developed a powerful set of gene-editing tools collectively called CRISPR. It is pronounced CHRIS-per and stands for—are you ready?—*Clustered Regularly Interspaced Short Palindromic Repeats.*

Got that?

Fortunately, you don't need to know exactly what it means to understand how CRISPR works, which conceptually is alarmingly simple.

Inspired by the extraordinary gene-editing abilities of a typical bacterium's immune system, of all things, CRISPR works this way:

1. For "eyes," it uses an organic molecule (ribonucleic acid, or RNA) to find the target gene's precise location along the long, twisted strand of DNA.
2. For "scissors," it uses special chemical enzymes (with names such as Cas9 and Cpf1) to snip both ends of the targeted gene.
3. Sensing damage, the wounded DNA molecule sets about to repair itself. At that very moment, CRISPR uses the aforesaid enzyme as "hands" to replace the excised gene with a new, ostensibly improved one.

CRISPR is currently being used to alter the genetic identity of plants, animals, and humans—which, understandably, has many cheerleaders hailing it as one of the greatest inventions of all time. Some of the hypesters are even predicting CRISPR will enable us to at last "cure diseases, curb world hunger, end pesticide use and save endangered species"[785]—in short, to create Garden of Eden 2.0.

Theoretically, it certainly has the potential to do that. But will it?

Will the creation story currently being written by genetic engineers turn out better than the biblical one? Or will the new storyline merely recapitulate the old one? Or worse.

Read on and judge for yourself.

# NOT OF THIS WORLD

*"Genetic engineering is to traditional
crossbreeding what the nuclear bomb
was to the sword."*

Andrew Kimbrell

One of my all-time favorite 1950s black-and-white horror movies is Howard Hawks' *The Thing from Another World*. It is set in the Antarctic and stars James Arness as an extraterrestrial, humanlike creature whose body is made of porous, unconnected tissue, with no arteries and no nerve endings.

"Just a minute, doctor," says Ned "Scotty" Scott, the lanky, bespectacled radio reporter covering the story. "Sounds like you're trying to describe a vegetable."

"I am," answers Dr. Stern.

"An intellectual carrot!" Scotty exclaims. "The mind boggles!"

Equally astonishing is how extremely far *real science* has come since 1951, when the movie was made. Yes, the idea of a human vegetable is still far-fetched, but as we're about to see, genetic engineers can now toy with plant DNA at will and are doing so to create fruits and veggies that are—well, not of this world.

## THE FIRST PLANT "THING"

The science-fiction–like adventure began in 1994, when the US Food and Drug Administration (FDA) approved the Flavr Savr tomato. I was at ABC News when the story broke, and I flew to the west coast for an exclusive taste test, courtesy of Calgene, the Davis, California, company that reportedly shelled out an estimated $20 million to create the unusual fruit.[786]

Calgene's scientists claimed to have successfully monkeyed with the tomato's aging gene in a way that curtailed the normal rotting process. I saw it as a forever-young tomato, even though it eventually does putrefy.

"The Flavr Savr will be a better-tasting tomato than the supermarkets carry most of the year," bragged Dan Wagster, Calgene's CFO. "How we price it has yet to be determined but consumers will have more value than ever before."[787]

I recall genuinely enjoying the tomato's robust flavor and saying so in a segment for *Good Morning America*. Over at NBC, Tom Brokaw declared, "The tomato stays riper, longer than the nonengineered variety, and they say it's tastier."[788] Mollie Davis, a food critic for *The New York Times*, averred that "the tomato was plump and juicy, far more supple than standard out-of-season tomatoes, possessing the deeper red hue and thinner skin of summer's field tomatoes."[789]

But the public didn't buy into the hype and ultimately gave Calgene's pioneering mutant tomato a big thumbs down. In 1997 it was withdrawn from the marketplace, and Calgene was purchased by Monsanto.

## THE LATEST PLANT "THINGS"

Today, there are five genetically engineered produce items we can buy at the grocery store: yellow summer squash, sweet corn, Rainbow papaya,[790] Innate potatoes,[791] and as of November 2017, Arctic apples.

The makers of Arctic apples—Okanagan Specialty Fruits (OSF) in Summerland, British Columbia—debuted their product in supermarkets around Oklahoma City.[792] I'm no longer at ABC News, but the storyline and accompanying hype are very much the same as they were for the Flavr Savr tomato more than twenty years ago.

This time, scientists have successfully figured out a way to silence the four genes that cause apple flesh to turn brown when cut, bruised, or aged. The brown is from melanin, the same coffee-colored chemical that tans our skin.[793]

"Overall, U.S. apple consumption is still far below what it was in the '80s and '90s, while obesity rates have sharply risen," says bioengineer Neal Carter, OSF's founder and president. Arctic apples "offer families an unparalleled eating experience that's also a wholesome, healthy snacking option."[794]

The sales pitch doesn't stop there. Turns out, an appalling number of apples are wasted each year—about 40% of the total. Many of them are partially eaten apples that people trash simply because of the flesh's off-putting russet color. "We strongly feel," says Carter, "that our non browning apples can help reduce that number."[795]

"But wait." As those infomercial TV barkers always say, "There's more!" Kids prefer eating apple pieces, rather than the whole fruit, says Carter, so "we have decided to initially offer Arctic apples in ready-to-eat bags of fresh cut apple slices." Besides, he adds, not missing a beat, "fresh slices also highlight our unique non-browning benefit and give consumers the convenience factor they desire to suit their busy lifestyles."[796]

Will people bite this time? Or will the Arctic apple go the way of the Flavr Savr tomato? Perhaps by the time you read this, the verdict will be in.

Whatever it turns out to be, companies other than Okanagan—which, by the way, is now owned by Intrexon, a huge, multinational biotech corporation[797]—have a huge stake in the outcome.

For instance, in December 2016, the FDA gave Del Monte Fresh Produce (DMFP) the go-ahead to sell its genetically engineered pink pineapple. "DMFP's new pineapple," the FDA explains, "has been genetically engineered to produce lower levels of the enzymes already in conventional pineapple that convert the pink pigment lycopene to the yellow pigment beta carotene. Lycopene is the pigment that makes tomatoes red and watermelons pink."[798]

DMFP is being hush-hush about the fanciful new fruit, but it has reportedly partnered with Dole to grow them in Costa Rica and Hawaii.[799] They're not in the supermarkets yet, but perhaps they will be by the time you read this.

## PLANT—ANIMAL "THINGS"

Beyond miracle tomatoes, apples, and pineapples is a burgeoning cornucopia of manmade plants and animals collectively called *genetically modified organisms*, or GMOs. Here are some examples of how far scientists are now venturing to tamper with plant DNA.

*Fish Genes*

In the 1990s scientists at DNA Plant Technology (DNAP) in Oakland, California, had a clever-sounding idea. Take the genes that protect arctic fish from freezing to death, insert them into a tomato's DNA, and—presto!—you'll have a frost-tolerant fruit.[800]

DNAP bioengineers followed through on the idea and successfully inserted antifreeze genes from a winter flounder into a tomato. They even received US Department of Agriculture (USDA) approval to plant the GMOs in a test field. In the end, however, DNAP did not take the fish-tomato to market.[801]

Some blame the company's abrupt decision to abort on the public's revulsion of the idea. "Believe me," says Jane Rissler, deputy director and senior staff scientist at the Union of Concerned scientists, "they would be doing it if people were not objecting to it."[802]

But DNAP disputes the accusation, saying the flounder's antifreeze genes simply failed to work well in the tomato plant. "The initial experiments showed insufficient technical effect to proceed with further experimentation and development," said Scott Thenell, then-director of regulatory affairs for DNAP, which is now out of business. "It simply was not worth pursuing."[803]

## Bacteria Genes

In 1987 scientists at Plant Genetic Systems in Ghent, Belgium, published a report announcing their creation of a new life form: a tobacco plant with genes from a common soil bacterium called *Bacillus thuringiensis* (Bt).[804] Their Bt-tobacco "thing" proved to be super-resistant to insects, because the bacterium's genes produce so-called Cry proteins deadly to all kinds of pests, though not to humans.[805]

Since then, scientists have inserted Bt-genes into the DNA of many crops we eat every day: corn, soy, cotton (as in cottonseed oil), potatoes, eggplants, and others.[806] And the strategy seems to be working. According to a 2016 study led by Kansas State University, the increasing number of farmers planting Bt-corn between 1998 and 2011 "is associated with a clearer decline in insecticide use." On average, a Bt-corn farmer uses 11% less insecticide than a conventional-corn farmer.[807]

But there is a significant fly in the ointment.

The dreaded corn rootworm beetle, for one, is developing a resistance to the Bt-gene's protection. "Resistance of corn rootworm to Bt-corn has been documented in parts of Iowa, Nebraska, and Illinois," reports the US Environmental Protection Agency (EPA). It adds "that other parts of the Corn Belt, where corn rootworm infestations are common and the use of Bt corn is high, are also at risk for resistance."[808]

The threat is a potential show-stopper, says a 2016 study led by Iowa State University and published in *Scientific Reports*. "The evolution of resistance and cross-resistance threaten the sustainability of genetically engineered crops that produce insecticidal toxins derived from the bacterium *Bacillus thuringiensis* (Bt)."[809]

## Human Genes

Genetic engineers are even inserting *human* genes into the DNA of many plants. These plant–human "things" are part of a larger experimental industry called *biopharming*. Biopharmers envision enslaving vast quantities of GM plants to mass-produce a dizzying array of industrial products—starches, fats, oils, waxes, plastics, etc.—as well as plant-grown medicines called *farmaceuticals*.

One big category of farmaceuticals comprises edible vaccines. These inoculations are delivered not by needle, but by eating, say, the leaves of a plant whose DNA has been re-engineered to produce antibodies.[810]

The biopharming craze began in the late 1980s but didn't take off right away because of some technical hurdles. Also, in 2002 the USDA and other federal regulators were spooked when experimental corn plants infused with animal genes contaminated food crops in Nebraska and Iowa. The errant experiment was run by ProdiGene, a biopharming company in College Station, Texas.[811]

There is conflicting information about which animal genes Prodi-Gene spliced into the corn's DNA, but the two foremost possibilities appear to be genes for producing a vaccine to fight diarrhea in pigs and genes to produce a controversial HIV vaccine.[812] In any case, the USDA fined ProdiGene $3 million and put an immediate kybosh on permitting any other farmaceutical field trials.[813]

Today, however, biopharming's fortunes are rebounding, with many predicting an explosive comeback.

The about-face began in 2012, when the FDA approved the world's first farmaceutical for humans: Elelyso, made by Protalix, an Israeli company, in partnership with Pfizer.[814] The drug—an enzyme called taliglucerase alfa—is produced by carrot plants modified with human genes. Elelyso is meant to treat Gaucher disease, which is most common among Ashkenazi Jews.[815]

That landmark achievement was followed by another in 2014, when an Ebola epidemic struck West Africa. Responding to the crisis, the FDA approved the emergency use of an experimental farmaceutical called ZMapp, a cocktail of antibodies produced by tobacco plants with humanlike genes. Miraculously, the drug—made by Mapp Biopharma-ceutical in San Diego, California—appeared to save the lives of seven people, including two infected American aid workers who were on the brink of dying.[816]

ZMapp has since been tested on seventy-two patients in a controlled study that yielded mixed results.[817] ZMapp appears to be completely safe, the study concludes, but is barely more effective than conventional Ebola treatments. "There's reason for optimism, but I think we can do better," says Erica Ollmann Saphire, a biologist at Scripps Research Institute. "I think science can do better."[818]

That stubborn belief—that science can do better—is certain to keep driving biopharming's current resurgence. As of this writing, companies worldwide are developing and testing scores of human–plant "things"— including transgenic rice, moss, alfalfa, and duckweed—designed to produce pharmaceuticals for everything from the flu to non-Hodgkin's Lymphoma.[819]

## TRUTH IN ADVERTISING

In the 1990s, as genetically modified foods started proliferating, I decided to do *Good Morning America* segments every year at Thanksgiving to show viewers what percentage of their traditional holiday meal was being replaced by GMOs. The speed of the takeover has been—to borrow Scotty's phrase—mind-boggling.

Today, according to the International Service for the Acquisition of Agri-Biotech Applications, more than eighteen million farmers in twenty-eight countries are growing genetically engineered crops. That's a one-hundred-fold increase in total acreage since 1996.[820]

According to the USDA, 94% of all soy beans, 92% of all corn, and 96% of all cotton grown in this country are GMOs.[821] That means an estimated 75% to 80% of everything we eat these days contains GMOs— from bread and tortillas to sodas and cooking oils.[822]

Most of the time we're not aware of it, because GMO food labels have not been required in the United States. The European Union began requiring GMO labels in 1997.[823]

In 2016 the US Congress finally did pass a labeling law; but it doesn't take effect until July 2018. At least, that's the plan. "We're a little behind to get this done by 2018," confesses Andrea Huberty, a USDA senior policy analyst. "We're still on track, but a little behind."[824]

Even when the law does take effect, it will not require GMO information to be printed on the food labels. Companies will have the option of enciphering the information in a QR code—which is readable only by scanning it with a smartphone—or merely referring consumers to a website or 1-800 phone number.

People wary of GMOs decry the subterfuge and are demanding more transparency. By contrast, farmers who believe GMOs are completely safe object even to this minimal kind of required labeling. "I don't think that it's the best bill that we could have," complains Richard Wilkins, president of the American Soybean Association, "but it's the best bill we could pass."[825]

## OLD-SCHOOL DANGERS

Any honest and thoughtful discussion about the safety of GMOs must begin with both sides acknowledging that even foods produced by Mother Nature aren't always safe to eat—among them, of course, many varieties of delicious-looking but poisonous mushrooms and berries. Even produce labeled "organic" or bred by using traditional artificial breeding techniques is not guaranteed to be safe.

Here are two examples that prove it.

*Killer Zucchini*

In early 2002 scores of New Zealanders reported getting sick after eating zucchini. The symptoms resembled those of stomach flu: cramps, diarrhea, and vomiting.

After investigating, botanists determined that most of the culprits were *organic* zucchinis, scrupulously farmed without pesticides or any other "unnatural" interventions. Moreover, they were grown from vintage seed varieties with a low tolerance for insect attacks—a significant point, it turned out, because aphids were unusually active during that year's growing season.

Without the aid of chemical pesticides, the organic zucchini plants were forced to fend off the aphid invasion entirely on their own. Which they did, by deploying a battalion of built-in, natural insecticides called *cucurbitacins*. The end result was that, by the time the zucchinis were ripe and ready to eat, the natural levels of cucurbitacins were so elevated they made people ill.

In retrospect, says William Rolleston, founder of New Zealand's Life Sciences Network and acting president of the World Farmers Organization, "the organic growers should have responded to the insect infestation with pesticides, or withdrawn their product from the market for food safety reasons." Obviously, he adds, just by "certifying food as organic does not make it safe."[826]

## Killer Potato

At the top of anybody's list of failed artificial breeding experiments is the notorious Lenape potato, named after a tribe of indigenous Americans also known as the Delaware Indians.[827]

In 1967 scientists at Pennsylvania State University and the Wise Potato Chip Company created the small, white, round spud by cross-breeding a wild Peruvian potato with a domesticated Delta Gold potato.[828] The Lenape boasted exceptional disease resistance and a low-sugar content, which produced potato chips and fries with a beautiful, golden brown color.

The prized Lenape was released with great fanfare. But very quickly, people reported feeling nauseous after eating it. Some even had convulsions.[829]

Scientists soon discovered the wonder-spud contained dangerously high levels of *solanine*, a natural chemical that accounts for the Lenape's remarkable disease resistance. Solanine—an alkaloid akin to nicotine, caffeine, and cocaine—is what turns a potato poisonously green if over-exposed to sunlight.[830] It also makes people sick.

The Lenape was withdrawn from the marketplace in 1970, but—spoiler alert—it is still being used for breeding stock. In fact, if you're a lover of potato chips or French fries, chances are you are eating the children and grandchildren of the ill-fated Lenape—including the Atlantic, Trent, Belchip, and Snowden varieties.[831]

## MY TAKE ON "THINGS"

Unlike The Thing in the movie, GMOs are not blood-thirsty monsters. But they do raise legitimate questions about what unchecked genetic engineering is capable of doing to the natural world—and our food supply.

In 2016 the National Academies of Sciences, Engineering, and Medicine sought to reassure the public that, for the most part, GM foods are safe to eat. They pointed out, for instance, that we and our livestock have been consuming GM crops for some years now without any noticeable ill effects.

I'm inclined to agree but with one important caveat. Our past experience with hugely popular products such as violet rays, radium, and other forms of radiation; smoking; DDT; lead paint; and cell phones teaches us it can take *many decades* for their harmful effects to become obvious.

What's more, our tinkering with the DNA of GM crops is still very much in its juvenile stages—creating relatively simple changes to their flavor, color, shelf life, and insect and herbicide resistance. The safety of GM foods will truly be put to the test in the coming years, as our genetic tampering gets more daring.

According to the aforementioned National Academies report, there are already some warning signs. For instance, herbicide-resistant crops are passing on their resistance to weeds, creating *superweeds*. That presents "a major agronomic problem," the study says, because it is forcing farmers to increase their use of potent, chemical herbicides to keep the superweeds in check.[832]

Genetic engineering is also creating powerful conflicts of interest. Case in point: Monsanto sells expensive, patented, herbicide-resistant crop seeds—including those in the popular, USDA-approved "Roundup Ready Xtend" line of GMOs. Monsanto also sells the patented herbicides to which their seeds are resistant: glyphosate (Roundup) and dicamba.[833] In other words, they're making money with both hands.

Monsanto's high-priced, proprietary products dominate agriculture so much now, farmers aren't able to freely trade seeds, as they once did. They fear Monsanto will sue them for patent infringements.

Scores of farmers are suing the company, claiming its death grip on the industry is forcing them out of business. These suffering farmers will soon "have to put up their equipment for auction," says Arkansas farmer Kenneth Qualls, a party to one class-action lawsuit, "and the people bidding on it [i.e., Monsanto's customers and allies] will be the ones who put them out of business."[834]

There's also concern about what the rising use of herbicides and herbicide-resistant crops is doing to the environment. For instance, to Monarch butterflies.

While at ABC News, I aired various segments on the winsome, fluttering, orange-and-black insects. I even traveled to the mountains of central Mexico, where Monarchs spend their winters, hibernating in pineapple-sized clusters that hang from the trees there like giant Christmas ornaments.

I learned Monarch butterflies are being devastated by wanton, illegal deforestation of those mountains—mainly, Mexicans cutting down trees for firewood. Changing weather patterns are also a problem, disrupting the Monarchs' migrations up and down North America.[835]

But a study by scientists in the United States and Mexico reveals there's another major threat: "the loss of breeding habitat in the United States due to the expansion of GM herbicide-resistant crops, with consequent loss of milkweed host plants."[836]

Monarchs lay their eggs in, and their larvae feed on, milkweed plants.[837] But the increased use of herbicide-resistant crops and herbicides is displacing and devastating the all-important milkweed plants through-out the United States.

From 1990 to 2010, the study says, the amount of milkweed and number of Monarchs in the Midwest alone fell by 58% and 81%, respectively.[838] During the same time, according to the USDA, the fraction of herbicide-resistant soybeans, corn, and cotton planted in the United States rose to 93%, 70%, and 78%, respectively.[839] So, as the acreage of herbicide-resistant GM crops soared, the population of Monarchs plummeted.

There's more disquieting news.

Many people defend our explosive use of GM crops by claiming that, acre-for-acre, they're more productive than natural crops and are, there-fore, our best hope for ending world hunger. But the National Academies report cautions against such hype, stating "there is no evidence from USDA data that they [GM crops] have substantially increased the rate at which U.S. agriculture is increasing yields."[840]

The general public, too, is cautious concerning the hype about GM foods. The Pew Research Center found 39% of Americans believe GM foods are *less healthy* than regular foods; only 10% believe GM foods are better for us.[841] Among Europeans, the GMO-wary percentages are even higher.[842]

## MY BIGGEST CONCERN

If you think only benighted laypeople are concerned about the safety of GM foods, you are mistaken. Well-informed experts are as well, citing the colossal difference between modern genetic engineering and age-old, traditional breeding.

"Such intervention [genetic engineering] must not be confused with previous intrusions upon the natural order of living organisms," warned the late Nobel Prize-winning Harvard biologist George Wald. "All such earlier procedures worked within single or closely related species. The hub of the new technology is to move genes back and forth, not only across species lines, but across any boundaries that now divide living organisms."[843]

People who do not grasp what Wald is saying here—that there is a *fundamental difference* between genetic engineering and conventional breeding—are my greatest concern. Consider, for example, American journalist Michael Specter—author of *Denialism: How Irrational Thinking Hinders Scientific Progress, Harms the Planet, and Threatens Our Lives*. "Genetic engineering," he claims, "is only one particularly powerful way to do what we have been doing for eleven thousand years."[844]

Whether it results from genuine ignorance, ideology, or a hidden agenda, such insouciance ignores the unique, perilous reality we are now facing. It risks lulling us into a false sense of safety.

As mere breeders, we could surely produce, say, a purple pumpkin. But it would take *years* of painstaking work to do so.

Today, armed with our knowledge of DNA and the powerful editing tools of genetic engineering—like CRISPR—we can, in principle, create a purple pumpkin in a matter of *days*, if not sooner. This sudden, god-like capacity portends a massive, unprecedented assault on the natural processes of heredity that have steered the trajectory of life since Earth's creation.

"With the insertion of the right snippets of DNA," says Harvard biologist E. O. Wilson, "new strains can be created that are variously cold-hardy, pest-proofed, perennial, fast-growing, highly nutritious, multipurpose, water-conservative, and more easily sowed and harvested. And compared with traditional breeding techniques, genetic engineering is all but instantaneous."[845]

Interestingly, Mother Nature appears to be a genetic engineer of sorts. On their own, genes have the uncanny ability to hop from one organism to another. It's called horizontal gene transfer (HGT), or *gene hopping*. One disputed study even goes so far as to claim that because of gene hopping, the human genome is infused with fully 145 genes from ancient plants, fungi, and other foreign microorganisms.[846]

Many hypesters use this information to defend genetic engineering. If gene splicing is good enough for nature, they declare, it should be good enough for us. Surely, there's nothing to fear.

The argument sounds eminently reasonable, but it is seriously flawed. Just because a process happens naturally doesn't mean it is safe for us to mimic—especially on a grand scale.

Rainfall, a thoroughly natural process, results from moisture accreting onto tiny airborne particles called cloud condensation nuclei (CCN), which ultimately produces raindrops. But that doesn't make it safe for us to seed clouds—to dust them artificially with CCNs on a grand scale. Can you imagine the disruption to Earth's weather patterns—and world politics—if Americans, say, decided to hog the atmosphere's moisture budget with a permanent, self-serving cloud-seeding initiative? It would spell disaster.

There is another critical difference between genetic engineering and traditional breeding.

As mere breeders, we're able to mate only certain species. We can mate a horse and a donkey to produce a mule, for example, or a lemon and a mandarin orange to produce a Meyer lemon.

But as genetic engineers, the possibilities are limitless. As we've seen, we can even "mate" a fish with a tomato.

Genetic engineering also gives us the power to *elevate recessive traits over dominant traits*—the very *opposite* of what nature does—on an unmatched scale. We have no way of knowing what effect such a willful affront will mean to the plant kingdom and ecosystem as a whole.

Ultimately, genetic engineering gives us the power to *destroy diversity*, by promoting the dominance of expensive, patented, GM crops over all other varieties. As far back as June 1988, more than two dozen African scientists warned that such a high-priced, domineering technology will not "help our farmers to produce the food that is needed in the 21st century. On the contrary, we think it will destroy the diversity, the local knowledge and the sustainable agricultural systems that our farmers have developed for millennia and that it will thus undermine our capacity to feed ourselves."[847]

So, no, genetic engineering is *not*, as Specter and others would have us believe, "only one particularly powerful way to do what we have been doing for eleven thousand years." It is an entirely different beast. As fundamentally different as mixing chemicals is from triggering a nuclear chain reaction.

Genetic engineering gives us the power to play God as never before, but without his wisdom or humility. If you doubt this, consider the hubris of the DNA pioneer James Watson. "If we don't play God," he boasts, "who will?"

In my opinion, our best hope of securing a great future and avoiding a grim fate lies with people who see with crystal clarity genetic engineering's unmatched power, promise, and peril. People such as Eva Novotny, former University of Cambridge astrophysicist and member of Scientists for Global Responsibility.

"There is no evidence that unlike species have ever [naturally] crossed during the billions of years that life has existed on earth," she explains. "We must not be so arrogant as to assume that we are more clever than Nature, lest we precipitate an irreversible chain of biological evolution that ends in catastrophe."[848]

Unless we comprehend the unprecedented power genetic engineering gives us to alter the genetic landscape, we will not use it with proper reverence and restraint. If we behave as if genetic engineering is just another form of conventional breeding, we risk moving blithely forward and visiting mayhem on the natural world—and our food supply.

That inexcusable ignorance is, above all, *The Thing* I fear most.

# THE LAND OF OZ

*"Men have become like gods. . . .*
*Science offers us total mastery over our environment."*

Sir Edmund Leach

In the famous 1939 movie *The Wizard of Oz*, Dorothy and her companions finally reach the Emerald City and promptly board a horse-drawn cab to go visit the great and powerful Oz.

"What kind of a horse is that?!" exclaims Dorothy, pointing at the carriage's bright purple horse. "I've never seen a horse like that before!"

"No, and never will again, I fancy!" replies the cheerful cabby. "He's the horse of a different color you've heard tell about."[849]

As the scene plays out, the horse's color changes from purple to red to yellow. There were no computer graphics back then, so to effectuate the dazzling special effect, filmmakers smeared grape-, cherry-, and lemon-flavored Jell-O pastes on three different horses.[850]

Remarkably, yesteryear's movie magic is today's scientific reality. No, to my knowledge, genetic engineers have not yet created a grape-, cherry-, or lemon-colored horse—or flying monkeys, for that matter, which are also in the movie. But, in principle, today's biologists do have the wherewithal to create such things, if they wish.

Indeed, biologists have the technical ability to manipulate every part of the DNA of every living organism on the planet—plants, animals, bacteria, you name it. Therefore, engineering something as outrageous as a purple horse or winged monkey raises not the question, "Can we?" but "Should we?"

The question is especially important because genetic engineers tinkering with DNA are like children playing with a new toy—one whose complexity is well beyond their full comprehension. As Craig Venter, the pioneering American geneticist who contributed greatly to the Human Genome Project, confesses: "My view of biology is: 'We don't know sh-t.'"[851]

Some observers downplay the potential risks of our newfound power and lack of understanding. "I suspect any worries about genetic engineering may be unnecessary," says British environmentalist James Lovelock. "Genetic mutations have always happened naturally, anyway."[852]

Others see considerable risk, owing not to the technology, nor our relative ignorance, but to our grasping motives. "Genetic engineering has never been about saving the world," warns Indian environmentalist Vandana Shiva, "it's about controlling the world."[853]

I agree with Shiva that our enemy is not science—which, at its best, is dedicated to opening our eyes to how the universe works—but our often willful, reckless use of it. Most of the time we mean well, but as we're about to see, even our noblest intentions reveal our audacious and risky power struggle with nature.

## LIFE FROM OZ

There are many laudable reasons genetic engineers wish to hack the DNA of animals. Doing so, they honestly believe, will enable them to find cures for deadly diseases, mass-produce medicines, and protect public health.

Case in point: Genetic engineers have transformed marmosets and cats into Oz-like creatures—animals of a different color, if you will—by supplementing their DNA with a *green fluorescent protein* gene from jelly fish. These *transgenic animals* glow with an eerie greenish color under ultraviolet light, so scientists can more clearly see how infectious diseases such as AIDS and Parkinson's spread throughout the body.[854]

But is exerting such an unprecedented degree of control over the animal kingdom a good thing? Do the ends truly justify the means?

Here are some other examples of what's happening in genetic engineering's land of Oz. Judge for yourself.

## Mental Health

"Globally," reports the World Health Organization, "more than 300 million people suffer from depression, the leading cause of disability." Moreover, it discloses, "depression and anxiety disorders cost the global economy $1 trillion US each year in lost productivity."[855]

In order for scientists to successfully tackle this scourge of depression and other mental afflictions, they must first learn how mammalian brains work. Many of them—like eager youngsters poking around the insides of a clock to see what makes it tick—are doing so with the help of genetic engineering.

To that end, neurobiologist Ivan de Araujo and colleagues at Yale University recently published the results of their landmark study of rat brains. Its primary goal was to learn precisely which neurons control the two prongs of a rat's all-important, predatory behavior: the *hunt* and the *kill*.

"Superior predatory skills led to the evolutionary triumph of jawed vertebrates," the researchers explain. "However, the mechanisms by which the vertebrate brain controls predation remain largely unknown."[856]

The Yale scientists discovered the hunt–kill neurons deep within the brain's emotional center—the *amygdala*. To confirm this, they infused the neurons with light-sensitive DNA from common algae, transforming them into Oz-like brain tissue. By hitting these neurons of a different color with blue laser light, the scientists were able to trigger the rats' hunt–kill behavior at will.

On command, the rats pounced on anything within sight, including bottle caps, wooden sticks, and live insects. "We'd turn the laser on and they'd jump on an object, hold it with their paws and intensively bite it as if they were trying to capture and kill it," reports de Araujo.[857]

The technique of using light to trigger certain behaviors is called optogenetics. In other labs, it's being used to force rats to run frantically in perfect circles or chug down milkshakes, and to force monkeys to abruptly swivel their gaze toward a designated target.[858]

Of course, this kind of basic research could one day result in effective treatments for depression and other severe mental afflictions, such as Parkinson's, Alzheimer's, and drug addiction. And that clearly is a great possibility.

But the research is already leading to mind-control technology that could one day be inflicted not just on defenseless rats and monkeys but on humans. Such a very real and present possibility is decidedly grim.

"This kind of direct mind control once belonged in science fiction," exults Ada Poon, an electrical engineer at Stanford University.[859] But, clearly, not anymore.

*Physical Health*

Motivated by wishes to "improve" nature—a conceit that presumes we humans know best—genetic engineers are meddling with the DNA of animals to make meats healthier and cheaper to produce. They also seek to turn animals into living factories to produce human medicines and human spare parts.[860]

Already approved by the FDA are:

- **Transgenic goats**—goats with human DNA—whose milk produces *antithrombin*. It's a blood-thinning protein used to treat people prone to lethal blood clotting in their legs and lungs.[861]

- **Transgenic chickens**—chickens with human DNA— whose eggs contain *sebelipase alfa*. It's a human enzyme that is the active ingredient in the drug Kanuma, marketed by Alexion, used to treat people who have trouble metabolizing fats. The malady leads to liver enlargement, fibrosis, cirrhosis, and cardiovascular disease.[862]

- **Transgenic salmon**—Atlantic salmon with genes from a Pacific salmon and an ocean pout—now being marketed with the brand name AquaAdvantage. The Oz-like fish grow with abnormal speed, reaching market weight twice as fast as normal salmon. AquaAdvantage debuted in Canadian grocery stores in 2017 but has yet to be sold in the United States, although it's expected they will be soon.[863]

Currently awaiting US government approval is a menagerie of other creatures from the land of Oz. They include: GM pigs with fleshier backsides and leaner pork chops; GM goats whose milk resists spoiling in warm weather; GM cows whose milk contains interferons, immunologic proteins that combat a wide range of human cancers and infections; and GM pigs to grow human-friendly hearts—a potential boon for thousands of Americans currently awaiting heart transplants.[864]

Muhammad Mohiuddin, a University of Maryland heart surgeon, for one, is elated about genetic engineering's power to refashion the animal kingdom in our own image. He sees animals with human DNA as wonderful vehicles for harvesting human-like, rejection-resistant spare parts, "eliminating the shortage of donor organs including hearts, livers, kidneys, intestine, as well as insulin-producing cells for treatment of diabetes."[865]

But some in the land of Oz are not content to stop there.

In 2016 the *MIT Technology Review* pulled back the curtain on renegade genetic engineers who are creating pig–human and sheep–human *chimeras*. Please note: whereas transgenic animals often have their DNA supplemented with *human genes*, these chimeras have actual *human cells*. Their vital organs aren't just human-friendly but actually human.

As I write this, the far-out creatures are quietly being created by at least three teams in California and Minnesota *against the wishes of and without any funding from the National Institutes of Health (NIH).* The scientists see themselves laboring for a noble cause (not to mention a Nobel Prize), namely, creating a limitless source of body parts for people in need of them.[866]

But others are concerned about the possible dangers.

What if, for example, the human cells in these illicit chimeras proliferate beyond their vital organs—to their brains, for instance? The resulting creatures, possibly capable of human thought, could no longer be considered just pigs.[867]

"We are not near the island of Dr. Moreau, but science moves fast," warns David Resnik, an ethicist at the NIH. "The specter of an intelligent mouse stuck in a laboratory somewhere screaming 'I want to get out' would be very troubling to people."[868]

Hiromitsu Nakauchi—a Stanford University biologist who is trying to make human–sheep chimeras—concedes such a thing could happen. "If the extent of human cells is 0.5 percent, it's very unlikely to get thinking pigs or standing sheep," he says. "But if it's large, like 40 percent, then we'd have to do something about that."[869]

## Public Health

I don't know of anyone who would plead for the life of a mosquito, given 750,000 people die annually of mosquito-borne diseases, such as malaria, dengue fever, yellow fever, Zika, and encephalitis. For us humans, the mosquito is by far the deadliest animal on the planet.[870]

There are roughly 2,500 species of the dreaded insect buzzing around every continent on Earth, except Antarctica. Over the years, science has invented increasingly clever and deadly methods for controlling the size and lethality of mosquito populations. Chemical sprays are among our oldest weapons of choice, but mosquitoes are developing a resistance to many of them.[871]

We've also tried sterilizing male mosquitos—for instance, by zapping them with radiation or infecting them with a disease. When the eunuchs are released *en masse* into the wild and mate with females, nothing comes of the unions. Thus, the overall mosquito population is knocked down.[872]

More recently, Oxitec—a company in Abington, England has focused on ridding the world of *Aedes aegypti* mosquitoes, responsible for spreading Zika, dengue fever, chikungunya, and yellow fever. Oxitec has figured out a way to infuse the DNA of male mosquitos with genes that cause its offspring to self-destruct before reaching adulthood. The lethal genes come from the *Escherichia coli* bacterium and herpes simplex 1 virus.[873]

The GM mosquitoes—code-named OX513A—also have supplemental genes that make them glow under certain light. Thus, scientists are better able to see the Oz-like insects' dispersal out in the field.[874]

The company has already released millions of OX513A in parts of Panama, Brazil, and the Cayman Islands—reportedly reducing wild *Aedes aegypti* populations by more than 90%, seemingly without any harmful effects to humans or the environment.[875] Despite the successes, however, Oxitec has been having a hard time persuading residents of the Florida Keys to allow trials in their backyards.

"I am going to ask you, beg you, plead with you, not to move forward with this," entreated resident Megan Hall at a 2016 board meeting of the Florida Keys Mosquito Control District (FKMCD).[876] And it's not just locals who are voicing concern. "DO NOT PERMIT!" writes a person from Lizuka, Japan. "Future ramifications are unknown and could be dangerous to the entire world."[877]

These critics point out, and Oxitec confirms, about 0.03% of the GM mosquitoes end up being females, which creates a risk—albeit tiny—of someone being bitten and infected with the OX513A's *E. coli*/herpes simplex 1 genes. Also, Oxitec's own lab research shows that about 4% of the mutant mosquitos' offspring do *not* self-destruct and actually live on to bite another day.[878]

Other skeptics point out that even if Oxitec's OX513As were ever to completely vanquish *Aedes aegypti*, other, equally dangerous mosquito species would inevitably flourish and fill the environmental vacuum—most notably, the Asian tiger mosquito, *Aedes albopictus*. "We've seen elsewhere . . . that when there's a vacant niche, *Aedes albopictus* will move in," notes University of Florida entomologist Phil Lounibos, "and it's a competent vector [i.e., effective transmitter] of the same viruses."[879]

All these concerns notwithstanding, a majority of Florida Keys residents and officials recently voted to allow the OX513A release. They point out that other possible remedies—such as spraying their neighborhoods with toxic pesticides—are hardly more desirable alternatives.

"I'm convinced that this technology is really needed in this industry," says Phil Goodman, the FKMCD's board chairman. "Because we are going to be dealing with these diseases for years to come, and coming out of the chemical industry, I can tell you there's nothing on the chemical side that's being developed for this."[880]

As of this writing, proposals to release millions of the OX513A mosquitos in the Florida Keys is stalled. The test has been approved by local officials and the EPA, but no actual *release site* has been identified and approved by residents themselves.

While we await the outcome, I can't help but remember what my friend and former Harvard colleague E. O. Wilson drummed into my head concerning the myriad, tiny, seemingly insignificant creatures that comprise the animal kingdom. "Remember, Michael," he cautioned me, "it's the little things that run the world."

Mosquitoes are thoroughly unpleasant, and it's hard to see what actual positive contribution they make to Earth's web of life. But any reasonable person must stop and wonder what damage we inflict on the exquisitely designed web every time we pull on even the most seemingly insignificant thread of it.

## LIFE FROM DEATH

Our age-old desire to control nature is especially evident in our current attempts to use genetic engineering to defy *death*. Here are two examples of what I mean.

### Cloning Animals

On Monday, February 24, 1997, I became the first TV correspondent to interview Ian Wilmut, the Scottish embryologist who caused a worldwide sensation by cloning a sheep. She was named Dolly after the ample-breasted country singer Dolly Parton, because Wilmut cloned her from an udder cell belonging to a six-year-old ewe.[881]

Wilmut achieved this seeming miracle by using a technique called *somatic cell nuclear transfer*.[882] SCNT was quite well known in science, but no one had ever used it to clone a mammal before—hence the screaming headlines and Dolly's picture on the cover of *Time* magazine.[883]

The name of the technique is admittedly a mouthful, but SCNT is actually easy to understand. Just bear in mind somatic cells are simply body cells, each of which houses an organism's DNA—the complete genetic recipe for making it from scratch.

Using SCNT, Wilmut extracted the DNA from the ewe's udder cell and inserted it into an egg whose own DNA had been removed. The egg—now thinking it was fertilized, because it had a full complement of DNA—was placed inside the womb of a surrogate mother. After a normal gestation period of 148 days, the mother gave birth to Dolly, an exact baby duplicate of the six-year-old ewe.[884]

In the months after Wilmut's bombshell achievement, a public debate raged about the ethics of cloning—especially when various shady groups came to light that vowed to use SCNT to clone *humans*. As ABC News's science correspondent, I was in the thick of covering the emotional back-and-forth. At one point—after infiltrating one particularly bizarre underground group—I even found myself at the center of the media circus.

In the years since, the commotion has died down considerably. But there's a chance it will ramp up again, because in January 2018, scientists from the Chinese Academy of Sciences, in Shanghai, announced they successfully used Wilmut's SCNT technique to clone two long-tailed macaque monkeys.[885]

Monkeys, like us, are primates—so the achievement hits close to home. Robert Lanza, who used the SCNT procedure to clone a gaur, calls the announcement "an impressive breakthrough, which overcomes the last major hurdle in the field."[886]

The Shanghai researchers insist they have no interest in cloning humans—but do concede their work could be used to further the controversial effort. "Monkeys are nonhuman primates that are evolutionarily close to humans," admits Muming Poo, a neuroscientist and cloning team member. But there "is no intention for us to apply this method to humans."[887]

While we wait to see where things go with primates, cloning mammals in general—including cows, horses, deer, mules, oxen, rabbits, and rats—has become a routine, money-making business.

One prominent cloning company—ViaGen Pets in Cedar Park, Texas—now offers ordinary people the opportunity to clone their favorite dog or cat. "People have a hard time wrapping their brain around that it is a real technology," says Melain Rodriguez, a client services manager for ViaGen.

But it is.

For $50,000, ViaGen technicians will remove some body cells from your favorite pooch, freeze them, if necessary, and then use SCNT to create a genetic copy of him. Cloning a cat costs half as much.[888]

Mike Hutchinson, a veterinarian at Animal General in Cranberry Township, Pennsylvania, had his toy poodle cloned.[889] He explains "even if you take a 16-year-old dog and you decide, 'Now I want to do it,' . . . [the cloning process] resets the clock. They [the dog's aged somatic cells] go back to young again, which is pretty unique."[890]

"It's not a reincarnation of their pet," Rodriguez is quick to clarify. "It's not that same pet born over again, but it's those same genetics and this . . . pet that they loved so much is somehow back in their life again."[891]

### Resurrecting Extinct Species

In the natural scheme of things, species come and go—sometimes explosively and in great numbers. Like the mass extinction 251 million years ago that wiped out more than 95% of all living species.[892] And the one 66 million years ago that snuffed out 76% of all species and 80% of all animal species, including the dinosaurs.[893]

Please note all five known mass extinctions happened long before we humans entered the picture, so we didn't cause them. The mass extinctions happened *naturally*, for reasons we honestly still do not understand.[894]

One certain take-away from the fossil record is this: nearly always, the extinction of one species creates opportunities for other species to rise and flourish. In that endless, unsentimental fashion, natural selection is always striving to maximize the planet's overall health—the way wise pruning promotes healthy branching patterns.

In a brazen attempt to interfere with this age-old natural process, genetic engineers are now attempting to *de-extinct* species long gone. In one greatly hyped initiative, scientists are trying to resurrect wooly mammoths, those ultra-photogenic, shaggy, pachyderms that began disappearing 10,000 years ago, with the rapid thawing of the little ice age.[895]

The idea was first made plausible in 2008, when a team led by Pennsylvania State University scientists announced it had successfully mapped a large portion of the wooly mammoth genome. The scientists did it by collecting DNA fragments from the hair of two frozen animal specimens.[896]

"If you want to bring a species back to life, the mammoth would be almost as dramatic as a dinosaur," says veteran science journalist Henry Nicholls. "It is a fair bet that a complete genome and closely related species would make it easier to pull a Crichton on a mammoth than on a dinosaur."[897] Michael Crichton wrote *Jurassic Park*, a novel wherein genetic engineers bring dinosaurs back to life—with disastrous consequences, I might add.[898]

Today, there are two prominent groups competing to make "Jurassic Park: Wooly Mammoth Edition" a reality. One of them is led by Harvard geneticist and media showman George Church.[899] The other, by the controversial South Korean biologist Hwang Woo-suk at Sooam Biotech, in collaboration with Russia's North-Eastern Federal University.[900]

Church's team has already implanted forty-five mammoth genes into the DNA of an Asian elephant—the mammoth's closest living relative—in hopes of one day creating a hairy, cold-resistant, mammoth-like elephant. One science writer suggests nicknaming the hypothetical, Oz-like creature a mammophant or elemoth.[901]

By contrast, Woo-suk's team is going for the whole enchilada, hoping to recover a completely intact mammoth genome from the frozen mammoth carcasses buried throughout the Siberian tundra. If successful, they will follow the same SCNT cloning procedure Ian Wilmut used to create Dolly. They will insert the mammoth DNA into a denucleated Asian elephant egg and then let it gestate inside an Asian elephant mother's womb for two years.[902]

Both teams face huge technical hurdles, but also questions about the wisdom of what they are doing. "The bigger question is: should we do it?" says cell biologist and science writer Helen Pilcher, author of *Bring Back the King: The New Science of De-Extinction.* "What would be the point?"[903]

Some argue we have a moral obligation to bring back species we unjustly killed off—for example, by our wanton destruction of their natural habitat. Others note that among all possible causes for the wooly mammoth's demise, humans are *not* at the top of the list. Instead, it's likely the hulking pachyderms—along with saber-toothed cats, ground sloths, Native American horses, camels, and much of the day's cold-hardy vegetation—simply failed to adapt to the precipitous global warming that happened 10,000 years ago.[904]

Of course, there's no denying we've caused or helped cause the extinction of countless other species. A 2014 study led by Duke University estimates that current annual rates of extinction are about 100 to 1,000 times higher than they were before we existed.[905] The implication, right or wrong, is we, along with climate change, are largely to blame.

Even if it's true, trying to de-extinct species is just further evidence of our stubborn desire to seize control of the natural world, without any real understanding of the possible consequences of doing so. I liken it to a well-meaning person without any carpentry skills offering to rebuild your house after tearing it down.

There are huge uncertainties associated with de-extinction. Could resurrected species survive in the modern world? Where exactly would we place them, given the original reasons for their disappearance might still exist, and might even be worse? And what about resurrected animals becoming pest species, foreign intruders that would disrupt today's ecology?

And that's not all.

A 2017 study led by Carleton University biologist Joseph Bennett and published in *Ecology & Evolution* found de-extinction efforts drain resources from much-needed conservation efforts racing to save *today's* precious, endangered species. "Even using the optimistic assumptions that resurrection of species is externally [i.e., not taxpayer] sponsored," the scientists report, the funding that inevitably would be needed to protect and preserve resurrected species "would lead to fewer extant species that could be conserved, suggesting net biodiversity loss."[906]

In other words, de-extinction efforts could make things worse, not better. According to the study, if we took the resources being used on today's flashy efforts to raise the dead and instead used them to rescue spectacular creatures currently teetering on the edge of the cliff—like black rhinos, mountain gorillas, and Sumatran tigers—we'd be able to save *two to eight times more* endangered species than we are doing right now.[907]

## LIFE FROM SCRATCH

As Hollywood publicists are wont to say about movie sequels, "If you liked how genetic engineers *mapped* plant and animal genomes, you're going to love how they're committed to *re-writing* them, with all sorts of customized changes thrown in."

Except this is not a movie pitch. It's happening.

As I explained at the start of this section, all DNA is constructed from just four main molecules—A, T, G, and C—which always pair up as AT and GC. With just these two pairings, it is possible to write an infinite number of genomes, or recipes for life. It's akin to being able to write an infinite number of books by using just the dots and dashes of Morse Code.

In 2010 a team led by geneticist Craig Venter successfully pieced together—DNA pair by DNA pair—the genome of a well-known bacterium. The scientists then implanted the freshly minted genome into the denucleated cell of a different bacterium. They named the resulting synthetized creature, appropriately enough, Synthia.[908]

That singular achievement not only made history, it inspired a bold, futuristic mindset. "Many scientists now believe that to truly understand our genetic blueprint, it is necessary to 'write' DNA . . . from scratch. Such an endeavor will require research and development on a grand scale."[909]

That mission statement comes from GP-Write (Genome Project-Write), a congress of scientists who aim to construct the genomes of every living organism on Earth *from scratch*, the way Venter's team did with Synthia. For these scientists, GP-Write is the logical sequel to GP-Read, the successful *mapping* of genomes, including the historic transcription of the human genome in 2003 [see FRANKENSTEIN: MEMORY LANE].

"We are going from reading our genetic code to the ability to write it," says Venter. "That gives us the hypothetical ability to do things never contemplated before."[910]

Besides replicating existing genomes, GP-Write scientists intend to *redesign* them and ultimately create entirely novel life forms, sprung entirely from the human imagination. As I said, the possibilities of writing genomes with just the AT and GC pairs are endless—including grand modifications to the natural, standard-issue human design.

"When people are thinking about the engineering or synthesis of the human genome, they immediately jump to a Brave New World of designer babies," says Nancy J. Kelley, a co-founder of GP-Write. "That's not where this project is going."[911]

But, of course, that is *precisely* where efforts like GP-Write are taking us, notwithstanding Kelley's astonishing naivete. As I explained in the last chapter, it is such self-inflicted blindness—scientists plunging ahead simply because they can, not necessarily because they should—that alarms me the most.

In *Heraclitean Fire: Sketches from a Life Before Nature*, Erwin Chargaff, the eminent Columbia University biochemist and DNA research pioneer, reflects on the two great discoveries he witnessed in his lifetime: nuclear energy and genetic engineering. Both resulted from hacking into once-sacred nuclei, he writes, "the nucleus of the atom, the nucleus of the cell. In both instances do I have the feeling that science has transgressed a barrier that should have remained inviolate."[912]

Before passing away in 2002, Chargaff was especially concerned genetic engineering was taking us in a grim and dangerous direction. Once we create a new life form, he said, "it will survive you and your children and your children's children." Such "an irreversible attack on the biosphere," he lamented, "is something so unheard of, so unthinkable to previous generations, that I could only wish that mine had not been guilty of it."[913]

# EYE OF THE BEHOLDER

*"With genetic engineering, we will be able to . . .
improve the human race."*

Stephen Hawking

Today, genetic engineers hacking the human body in unprecedented ways claim their efforts will ultimately improve the species. But will they, really?

I'm reminded of the *Twilight Zone* episode, in which a team of doctors has repeatedly tried improving the looks of Janet Tyler, a young woman whose bandaged face we're told is hideously disfigured. After giving it one last go, the surgeon, anesthetist, and nurses gather expectantly around Janet's bed, as the bandages slowly come off.[914]

"All right, Miss Tyler," the surgeon says kindly. "Now here comes the last of it. I wish you every good luck."

A moment later, the nurses gasp.

The doctor, dropping the scissors, recoils in horror. "No change!" he exclaims. "No change at all!"

Instantly—in a classic *Twilight Zone* twist—we see Janet Tyler's beautiful face and the frantic, grotesque visages of the surgeon, anesthetist, and nurses. They look like mutant pigs, large snouts and all.

Today, biohacking—the use of genetic engineering to "improve" not just our external appearance, but the entire human design—is raising some important ethical questions. What is biohacking's ultimate vision of perfection, exactly? And who decides what constitutes imperfection?

"The best time to have these conversations about a new technology is right before it becomes plausible," says Hank Greely, a bioethicist at Stanford. "Now is the time to talk about it."[915]

Iceland illustrates perfectly the frightful ethical problems with seeking human perfection. Since the start of the millennium, nearly 100% of that nation's women have chosen to abort fetuses prenatal testing indicates have Down syndrome, a genetic condition caused by an extra chromosome.[916] Today, only about two such allegedly imperfect Icelandic children slip through the cracks and are born every year.

"We don't look at abortion as a murder," explains Helga Sol Olafsdottir, a counselor at Reykjavik's Landspitali University Hospital. "We look at it as a thing that we ended. We ended a possible life that may have had a huge complication . . . preventing suffering for the child and for the family. And I think that is more right than seeing it as a murder. . . ."[917]

Think what you will about abortion, Olafsdottir's rationalization doesn't square with the facts. In a 2011 study led by Massachusetts' famed Children's Hospital Boston, scientists found "nearly 99% of people with Down syndrome indicated that they were happy with their lives; 97% liked who they are; and 96% liked how they look. Nearly 99% of people with Down syndrome expressed love for their families, and 97% liked their brothers and sisters."[918]

Adults with Down syndrome routinely lead productive lives, even earning college degrees, becoming artists, running companies, and winning political offices.[919] Karen Gaffney, a woman with Down syndrome, swam the English Channel, escaped Alcatraz *sixteen times*, and delivered a rousing TED talk titled, "All Lives Matter."[920]

## DER ÜBERMENSCH

Iceland's disdain for supposedly imperfect persons is not anything new, as illustrated by Plato's report in the fourth century BC about a physician named Asclepius. "But when it came to people whose bodies were permeated with disease, he did not attempt to extend their useless lives . . . and have them producing children who would probably be just like them," Plato writes in *The Republic*, "since treating them did no good either for the patients themselves or for the state."[921]

In 1866 that same brutal worldview got an unexpected boost from Charles Darwin's notion that nature is constantly improving itself by selecting strong heritable traits. "As natural selection works solely by and for the good of each being," Darwin wrote in *On the Origin of Species by Means of Natural Selection*, "all corporeal and mental endowments will tend to progress towards perfection."[922]

The eminent English scholar Francis Galton, Darwin's half-cousin, decided *civilized societies* could achieve perfection in the very same way. "[A]s it is easy . . . to obtain by careful selection a permanent breed of dogs or horses gifted with peculiar powers of running," Galton explained in his 1869 book *Hereditary Genius*, "so it would be quite practicable to produce a highly-gifted race of men by judicious marriages during several consecutive generations."[923]

To describe his scheme for creating the perfect society, Galton coined the term *eugenics*, from the Greek words for "well-born" or "good race/stock."[924] He also began an aggressive campaign for the procreation of admirable traits—as exemplified, he said, by the British aristocracy—*and* against the procreation of feeble traits, which he ascribed to the indigent, promiscuous, and criminal classes.[925]

His eugenics campaign immediately won the endorsement of his famous cousin. "Though I see so much difficulty [actually implementing such a plan], the object seems a grand one," Darwin wrote to Galton, "and you have pointed out the sole feasible, yet I fear utopian, plan of procedure in improving the human race."[926]

Very quickly, many other intellectuals of the day hailed Galton's plan for perfecting the human race—including George Bernard Shaw, H. G. Wells, and Bertrand Russell. "The way of nature has always been to slay the hindmost," observed Wells. "It is in the sterilization of failures . . . that the possibility of an improvement of the human stock lies."[927]

With the approach of the twentieth century, Galton's eugenics movement soared across the Atlantic and promptly won the enthusiastic support of America's intellectuals. They included faculty members from universities nationwide, and even Harvard University's fabled presidents Charles Elliot and A. Lawrence Lowell.

Soon, scholarly publications were rife with pro-eugenics scientific studies maligning any and all allegedly inferior groups of humans. Typical of them was a 1906 scientific article in the *American Journal of Anatomy* titled, "Some Racial Peculiarities of the Negro Brain."[928]

Between 1907 and 1931—the height of the American eugenics movement—thirty states passed laws to do what H. G. Wells declared was needed: force the sterilization of people considered undesirable. Many of the laws were successfully struck down in court. But in 1927—in a landmark case involving Virginia's proposed sterilization of Carrie Buck, an unwed mother with a moron-level IQ—the US Supreme Court decided in favor of the state, eight to one.

"It is better for the world, if instead of waiting to execute degenerate offspring for crime, or to let them starve for their imbecility," wrote Supreme Court Justice Oliver Wendell Holmes Jr., "society can prevent those who are manifestly unfit from continuing their kind." Referencing Virginia's determination that Carrie's mom *and* Carrie's infant daughter were also feebleminded, Holmes concluded by saying, "three generations of imbeciles are enough."[929]

During those heady days of eugenics, a young activist named Margaret Sanger founded the American Birth Control League—a precursor of today's Planned Parenthood Federation of America—*and* the Birth Control Clinical Research Bureau (BCCRB). Situated in New York City, the BCCRB offered women information about and prescriptions for various kinds of contraceptives, including diaphragms and jelly, condoms, and the rhythm method.[930]

Sanger couched her activism in feminist terms, for which today's political progressives celebrate her. Yet in 1934 Sanger authored the American Baby Code, in which she proposed giving the government enormous sway over women. "No woman shall have the legal right to bear a child . . .," she wrote in Article 4 of the code, "without a permit for parenthood."[931]

Sanger advocated such Draconian edicts because she was an unblinking eugenicist. Indeed, she was a more blunt-speaking eugenicist than even H. G. Wells, with whom she had a well-publicized affair.[932]

"Every . . . malformed child, every congenitally tainted human being brought into this world is of infinite importance to that poor individual; but it is of scarcely less importance to the rest of us and to all of our children who must pay in one way or another for these biological and racial mistakes," she wrote in her 1922 best-seller *The Pivot of Civilization*. "We are paying for and even submitting to the dictates of an ever increasing, unceasingly spawning class of human beings who never should have been born at all . . ."[933]

Contrary to the assertions of her modern-day apologists, Sanger did not see eugenics and birth control as separate social campaigns. "[T]he campaign for Birth Control is not merely of eugenic value," she wrote in *The Eugenic Value of Birth Control Propaganda*, "but is practically identical in ideal, with the final aims of Eugenics."[934]

If there was still any doubt in anyone's mind about Sanger's final intentions concerning birth control, she made them perfectly clear in the book *Woman and the New Race*. "Birth control itself, often denounced as a violation of natural law, is nothing more or less than the facilitation of the process of weeding out the unfit, of preventing the birth of defectives or of those who will become defectives," she said.[935] "Birth control must lead ultimately to a cleaner race."[936]

Outspoken, public support for the eugenics movement plummeted in the 1940s, after the Nazis took it to an extreme, by exterminating more than six million Jews, Romanies, Jehovah's Witnesses, criminals, disabled persons, and other supposed undesirables.[937]

"Darwinism by itself did not produce the Holocaust," observes historian Richard Weikart at California State University, Stanislaus, "but without Darwinism . . . neither Hitler nor his Nazi followers would have had the necessary scientific underpinnings to convince themselves and their collaborators that one of the world's greatest atrocities was really morally praiseworthy."[938]

According to Hitler's economist and confidant Otto Wagener, the Führer drew inspiration for the unspeakable atrocity from American eugenicists. "I have studied with great interest," Hitler reportedly said, "the laws of several American states concerning prevention of reproduction by people whose progeny, in all probability, be of no value or be injurious to the racial stock."[939]

Despite this sordid history, the basic philosophy behind eugenics— that some people don't deserve to live and should be prevented from existing—is still very much alive today, as we see in Iceland. In fact, thanks to genetic engineering, eugenics (though it is not called that) is now more popular than ever.

## NEW AND IMPROVED

For Johnjoe McFadden, a molecular geneticist at the University of Surrey, the possibilities of using genetic engineering to improve the human lot are endless and exciting. "Genetic editing is like playing God," he declares, "and what's wrong with that?"[940]

Here are two prime examples of what's in store for us. Decide for yourself how to reply to McFadden's question.

### Customized Offspring

On August 2, 2017, a medical research team led by the Oregon Health & Science University (OHSU) in Portland announced it had done something historic: it successfully biohacked dozens of human embryos plagued with a defective MYBPC3 gene that causes *hypertrophic car-*

*diomyopathy*, a congenital heart condition that afflicts about one in 500 people. Using the popular gene-editing tool CRISPR-Cas9 [see FRAN-KENSTEIN: MEMORY LANE], the scientists snipped off the defective MYBPC3 gene and replaced it with a normal one.[941]

All told, the scientists hacked fifty-eight defective embryos, of which forty-two were successfully corrected—an impressive 72.4% success rate. That bests by far previous attempts at hacking human embryos by other groups in China,[942] Sweden,[943] and the United Kingdom.[944]

The scientists behind this remarkable achievement say their technique could one day let us "fix defective" human embryos, instead of aborting them. "Discarding half the embryos is morally wrong," says Shoukhrat Mitalipov, a Kazakhstani-born biologist and the director of OHSU's Center for Embryonic Cell and Gene Therapy. "We need to be more proactive."[945]

Beyond such inevitable, utopian hype, many scientists and bioethicists are deeply worried that tinkering with the DNA of a human embryo to fix one disease could inadvertently create other genetic diseases. Such unimaginable mutant afflictions could forever infect the human gene pool.

"When you're editing the genes of human embryos, that means you're changing the genes of every cell in the bodies of every offspring, every future generation of that human being," notes Marcy Darnovsky, director of the Center for Genetics and Society in Berkeley, California. "So these are permanent and probably irreversible changes that we just don't know what they would mean."[946]

There's also a great concern embryo-hacking research will inexorably lead to a booming business in designer babies. "In a world dominated by competition, parents understandably want to give their kids every advantage," says the eminent NYU bioethicist Arthur Caplan. "The most likely way for eugenics to enter into our lives is through the front door, as nervous parents . . . will fall over one another to be first to give Junior a better set of genes."[947]

Being able to create designer babies, says Darnovsky, raises the specter of a grim future indeed. "If we're going to be producing genetically modified babies, we are all too likely to find ourselves in a world where those babies are perceived to be biologically superior," she warns. "And then we're in a world of genetic haves and have-nots. That could lead to all sorts of social disasters. It's not a world I want to live in."[948]

Yet, thanks to the prospects of a new, profit-making industry, such a world is fast approaching. Today, laying the groundwork for it, commercial startups all over the world are investing hundreds of millions of dollars in CRISPR technology.[949]

"When we spliced the profit gene into academic culture, we created a new organism: the recombinant university," laments the American Nobel Prize-winning biochemist Paul Berg. "The rule in academe used to be 'publish or perish.' Now bioscientists have an alternative: 'patent and profit.'"[950]

In February 2017, twenty-two bioscientific leaders commissioned by the National Academies of Sciences, Engineering, and Medicine (NASEM) concurred with all the above-stated concerns. But it didn't stop them from blessing further human embryo-hacking research, albeit subject to a long list of toothless caveats. "Heritable germline genome-editing trials must be approached with caution," the NASEM report concluded, "but caution does not mean they must be prohibited."

So much for science policing itself, says George Annas, the distinguished Boston University bioethicist. "The scientists are saying this is all a question of risk-benefit analysis, versus saying, 'No, it's just wrong to do,'" he says. "It's like torture: some people think we should never do it, other people say, 'No, no, if it works, then it's okay.' Designer babies is a lot like that."[951]

Presently, in the United States and the European Union, there are laws limiting or outright prohibiting tampering with the DNA of human embryos.[952] But let's be honest, nothing can possibly stop renegade scientists from doing whatever they want. As Nicholas Rescher, American philosopher at the University of Pittsburgh, observes, "Any attempt to set 'limits' to science . . . is destined to come to grief."[953]

The aforesaid team of OHSU-led, embryo-hacking scientists, for example, got around stiff US regulations by using institutional and private monies, not government funding.[954] Moreover, its leader, Shoukhrat Mitalipov, is a well-known maverick.

In 2007 he cloned monkey embryos[955] and in 2013 produced stem cells by cloning human embryos.[956] Now, despite professing opposition to designer babies, Mitalipov is making it clear no one and nothing will keep him from his goal of producing full-term, genetically engineered human babies—even if it means relocating to a country where it is allowed. "We'll push the boundaries," he vows.[957]

And he's not alone in jumping for joy over the possibility of creating genetically engineered human babies.

"This is exciting," avers physiologist Sakthivel Sadayappan of the University of Cincinnati. "This is the future."[958]

"The scientists are out of control," warns bioethicist George Annas. "They want to control nature, but they can't control themselves."[959]

## Customized Parents

John Zhang, founder of the New Hope Fertility Clinic in New York City, exemplifies what Annas fears most.

In 2003—when Zhang was at Sun Yat-Sen University of Medical Science in Guangzhou, China—he and some colleagues there announced the invention of novel techniques for creating what the media dubbed *three-parent babies*.[960] The procedure was widely slammed as unethical, but Zhang's motivation seemed reasonable: give women with defective mitochondrial DNA the chance to have healthy offspring.

Mitochondrial DNA—*mtDNA*—is made of the same stuff as ordinary nuclear DNA, *nDNA*. But whereas nDNA is linear and dwells within a cell's nucleus, mtDNA comes in small loops and lives in tiny islands called mitochondria outside the nucleus.[961]

Moreover, nDNA controls our entire being, with about 20,000 to 25,000 genes [see FRANKENSTEIN: MEMORY LANE].[962] By contrast, mtDNA controls only the machinery within our cells that converts food into energy, using just thirty-seven genes.[963]

If a woman's eggs have normal nDNA but defective mtDNA, her offspring can inherit certain terrible diseases. A man's mtDNA is never passed on to the next generation, because it is obliterated during the fertilization process, so it is of no concern here.[964]

In 2016 Zhang announced he used one of his three-parent-baby techniques on a thirty-six-year-old Jordanian woman whose defective mtDNA is known to cause Leigh Syndrome. It's a rare fatal disorder that attacks the central nervous system, usually starting at a very young age.[965]

Zhang conducted the headline-making experiment in Mexico, because it is not allowed in the United States. Here's how it worked.

Zhang collected eggs from both the Jordanian patient *and* a female donor. Next, he removed the nDNA from the donor eggs and replaced it with nDNA from the patient's eggs. Finally, he fertilized the engineered eggs with sperm from the Jordanian patient's husband. The resulting fertilized eggs, therefore, had DNA from three different people: healthy nDNA from the patient, healthy mtDNA from the female donor, and healthy nDNA from the patient's husband.

All told, the controversial procedure netted *five* three-parent fertilized eggs. One of them was viable enough to be implanted in the Jordanian woman's womb.

Nine months later, in April 2016, she gave birth to a seemingly healthy boy. It was a stunning achievement, given her defective mtDNA had hitherto caused four miscarriages and the death of two children, eight months old and six years old.[966]

"It's exciting news," says Bert Smeets, a biochemist at Maastricht University in the Netherlands.[967] "This is great news and a huge deal," agrees Dusko Ilic, a biologist at London's King's College. "It's revolutionary."[968]

This might have been the happy ending to John Zhang's story. But there's more.

In 2016 Zhang quietly started a for-profit company called Darwin Life, offering mitochondrial replacement therapy (MRT) to women forty-two to forty-seven years old for $80,000 to $120,000 a pop.[969] According to a 2015 report, American women over forty-two who resort to IVF get pregnant only 3.2% of the time.[970] Zhang's MRT is based on unsubstantiated speculation that faulty mtDNA is what causes older women to have difficulties conceiving.

According to the US Securities and Exchange Commission, Zhang has raised $1 million in initial funding for Darwin Life. But he hasn't said from whom.[971]

There are other worrying facts.

In 2016 Zhang and colleagues published a paper that contains this explicit disclaimer: "Declaration: The authors report no financial or commercial conflicts of interest."[972] There is no mention whatsoever of Darwin Life.

To skirt US regulations severely restricting the biohacking of human embryos, Zhang announced Darwin Life would create three-parent eggs in his New York facility but perform the actual surgery in his facility in Mexico, where "there are no rules." Says he, "For now, our nuclear transfer technique is very much like an iPhone that's designed in California and assembled in China."[973]

Even scientists who hailed Zhang's successful experiment with the Jordanian woman say his commercial venture is troubling because MRT technology is still way too untested. Despite Zhang's efforts, for instance, the Jordanian baby is still infected with a few percent of his mother's defective mtDNA.[974] Scientists say it's conceivable, as the boy matures, these defective seeds could overrun his normal mtDNA, ultimately resulting in his falling victim to the dreaded Leigh Syndrome.

Scientists are also worried about the vast uncertainties of mixing and matching DNAs from three parents. Genetic incompatibilities among them, they caution, could result in completely unforeseen complications as the child grows up.

We just don't know.

"The number of issues that are still unresolved—it's just staggering," says David Clancy, an expert in mitochondrial biology at England's Lancaster University.[975]

In a sternly worded letter dated August 4, 2017, the FDA reprimanded Zhang for his sketchy behavior and dubious motives with

Darwin Life. "Please be advised that you are using MRT to form a genetically modified embryo, which is subject to FDA's regulations," the letter says. "The genetically modified embryo that you formed using MRT does not meet all the criteria . . . Nor is exportation [e.g., to Mexico] permitted . . ."[976]

Today, if you type in the URL www.darwinlife.com, you're sent to a flashy website with this disclaimer:

> "The information contained on this site is intended for informational purposes only. Please be advised that MRT and IVF using MRT-modified oocytes is not performed in the United States, and has not been reviewed by the U.S. Food and Drug Administration. Also . . . we cannot guarantee procedures will result in successful outcomes and healthy children. . . . and the company will not conduct any clinical work in USA [sic] for the foreseeable future."[977]

In the meantime, others are speeding ahead with their own versions of MRT. In January 2017, Valery Zukin, director of the Nadiya Clinic of Reproductive Medicine in Kiev, announced his team used a technique similar to Zhang's on an infertile woman. The result, he claims, is a healthy baby girl.[978] And in March 2017, after years of heated debate, England's Human Fertilisation and Embryology Authority gave doctors at Newcastle Fertility Centre the go-ahead to use three-parent MRT on patients with *severely defective* mtDNA.[979]

As for Zhang, the criticism from fellow scientists and his run-in with US authorities is unlikely to stop him. He is clearly a man on a mission, which goes well beyond aiding infertile older women and women with defective mtDNA. It's a mission redolent of the passionate eugenics campaigns waged by the likes of Galton, Darwin, Wells, and Sanger.

Zhang says he is determined to continue expanding his technical repertoire, so one day parents can have perfect children. From the color of a child's hair and eyes to their IQ and athleticism—everything will be on Darwin Life's biohacking menu of the future.

"Everything we do is a step toward designer babies," Zhang says candidly about his ultimate scientific and commercial goals. "With nuclear transfer and gene editing together, you can really do anything you want."[980]

## CLOSING THOUGHTS

Jef Bocke, a geneticist at New York University and co-founder of GP-Write, is among those who don't get why people are concerned about science reconfiguring the DNA of all living organisms. "Unless they subsist exclusively on fruits, nuts, and fish," he says, "there is about a 100 percent chance they are enjoying the meddling done by our genetically oriented forebears who did selective breeding."[981]

As I have stated elsewhere in this section, this nonchalance—this misguided belief that nothing has really changed and genetic engineering is essentially no different than selective breeding poses a far greater danger to our future than the technology itself.

To be sure, since the dawn of life, we have significantly reshaped it, by systematically favoring certain traits and disfavoring others. But as I have attempted to explain in these chapters, genetic engineering is a completely different beast.

As mere breeders, we are like gamblers with clever systems for winning. We game the odds in our favor by betting in certain, calculated ways. But in the end, the laws of chance remain in control; they call the shots and determine our successes and failures.

As genetic engineers, however, we can now actually rig the decks of cards and wheels of fortune, to make them do exactly what we want. No longer at the mercy of chance, *we* now call the shots. *We* make DNA dance to our tune.

"I think DNA is going to be the most important material of the 21st century," says Emily Leproust. She is the CEO of Twist Bioscience, a popular San Francisco-based manufacturer of customized DNA strands. "The last century was about computers, and now we are entering an era of biology."[982]

This heady, unprecedented power will enable us to create a truly designer world. A vain reality that conforms perfectly with our egocentric needs and desires—just as we see happening in Iceland.

Icelanders have successfully created a world scrubbed of people with Down syndrome. But at what cost to their nation's heart and soul? And, speaking of soul, is it a mere coincidence Iceland is fiercely atheistic?[983] A magazine article recently boasted 0.0% of Icelanders under twenty-five believe in God.[984]

In 2009 Icelander Thordis Ingadottir inadvertently gave birth to Agusta, a Down syndrome child—one of only three born that year. Today, despite Iceland's eugenics-like culture, Ingadottir has high hopes for her daughter's happiness. "I will hope that she will be fully integrated on her own terms in this society," she says. "That's my dream."[985]

But, honestly, how can Agusta ever feel at home in a nation hell-bent on eliminating her kind? Think back to the plight of young Janet Tyler in that *Twilight Zone* episode. After her last failed surgery, the pig-faced people ostracized Tyler to a village of other "misfits," just like her.

"The problem with eugenics and genetic engineering," says Michael J. Sandel, political philosopher at Harvard University, "is that they represent a one-sided triumph of willfulness over giftedness, of dominion over reverence, of molding over beholding."[986]

Iceland, of course, is entitled to determine its own future. But is it reasonable to say its feticidal practices are *improving* civilized society or *perfecting* the human race? Is Iceland superior to nations that value and nurture people with special challenges and aptitudes?

"Christians need to take the lead in educating people that children are gifts," the late Charles Colson believed, "as my autistic grandson most surely is. By going down the path we're currently on, we might one day get rid of genetic diseases, but only at the cost of our own humanity."[987]

Lured by the conceit in thinking we can improve on billions of years of natural selection—or, if you prefer, on God's creation, or both—will we truly end up creating a better world? And if so, for whom will it be better?

Clearly, not for the Agustas of the world. That's why her mother, Thordis, is now asking: "What kind of society do you want to live in?"[988]

There's an even deeper question we must answer, says Yuval Noah Harari, an Israeli historian at Hebrew University of Jerusalem. An urgent question crying out to us, as we race to bring humans and all creatures great and small under our complete control. As we rush pell-mell to redesign life in our own selfish image.

"The real question facing us is not 'What do we want to become?'" he says, "but 'What do we want to want?' Those who are not spooked by this question probably haven't given it enough thought."[989]

I completely agree—and hope this book has, at the very least, given you, my dear reader, plenty of food for thought.

# CONCLUSION

# SCIENCE, SACREDNESS, AND SELF-DESTRUCTION

*"Our scientific power has outrun our spiritual power.*

*We have guided missiles and misguided men."*

Martin Luther King Jr.

Throughout this book, I have stated that my greatest apprehensions about the future are not the audacious technologies we're creating. My biggest fear is the hypester who downplays their unprecedented potential to destroy us.

For such a person, today's innovations—computers, the web, social media, surveillance cameras, tracking software, AI, killer robots, self-driving cars, sexbots, cyborgs, GMOs, manmade DNA, cloning, designer babies—aren't any more or less revolutionary than the myriad inventions scientists and engineers have been dishing out for centuries. We'll survive them all, the hypester shrugs dismissively, just as we've survived trains, planes, automobiles, rockets, and nuclear energy.

For the record, I agree it's likely at least some of us will survive the oncoming future. But I hope we're aiming for more than just staying alive.

In any case, this time, I believe, our situation is different. Very different. And it's not just because—as I've explained in this book—we can now force nature to do our bidding, more than at any other time in history.

No, this time is unlike all other times because of our drastically altered estimations of science and ourselves. Today, in a striking reversal of fortunes, *we see science as sacred and ourselves as not.*

That singular juxtaposition, I believe, threatens to create a future bristling with unique risks. For as American scholar John Schaar once noted, "The future is not some place we are going, but one we are creating. The paths are not to be found but made. And the activity of making them changes both the maker and the destination."[990]

If we have any real hope of seizing the great and avoiding the grim possibilities of our technology, I believe we need the guidance of wisdom that transcends our own self-centered lights. Wisdom well beyond the dubious, overhyped, utopian visions of science, such as the eugenics movements of the past and present. Imperishable wisdom offered to us by the world's religions—including my own, Christianity.

## OUR UNIQUE DUAL-MINDEDNESS

Besides being scientifically minded, we humans are without question spiritually minded. The two predilections are imprinted on our DNA, as surely as our hair and skin colors.

It's possible we might never be able to identify the exact sequences of A-C-T-G that code for this unparalleled dual-mindedness of ours. But the evidence for its existence is plainly seen in our stunning spiritual and scientific legacies—among them, millions of houses of worship blanketing the earth and scores of human footprints on the moon.

That said, in my lifetime, religious thinking has taken a beating from the champions of secularity in academia and the mainstream media. One of their favorite ploys is to accuse religion of being the top cause of human violence in history, which is patently untrue.

In the three-volume *Encyclopedia of Wars*, authors Charles Phillips and Alan Axelrod explain that of the 1,763 well-documented wars in human history, only 123 involved a religious cause. Sixty-six were fought in the name of Islam, the remainder in the name of all other religions. All told, religious strife has caused only about 6.98% of all wars and fewer than 2 percent of total wartime casualties.[991]

Despite this, witty atheists with a theatrical flair, such as comedian Bill Maher or Oxford University's evolutionary biologist Richard Dawkins, have turned religion-bashing into a lucrative spectator sport, today's version of tossing Christians to the lions.

"To be fair, much of the Bible is not systematically evil but just plain weird, as you would expect of a chaotically cobbled-together anthology of disjointed documents . . ." claims Dawkins.[992] He's been photographed wearing a T-shirt that reads Religion: Together We Can Find The Cure.[993]

But in spite of a daily torrent of such insults, glib denunciations, and show-boating, religious thought remains a central and enduring feature of the human experience. In fact, as we're about to see, far from being a disease in need of curing, religious thought is one of the few things uniting us all, both East and West—no small feat.

## THE RELIGIOUS WORLDVIEW

Despite their very real differences, the world's major religions agree on this: the cosmos has essential traits imperceptible to our keenest senses or finest instruments. These include spiritual agents conspicuously influencing our lives as much, if not more so, than, say, gravity—itself an invisible, yet mighty force.

The world's religions also agree on something else: spiritual phenomena are governed by *laws*—exactly the way physical phenomena obey, say, Newton's three laws of mechanics. Here are some examples of what I mean.

- **Hindus** believe in the *law of karma*, which is redolent of Newton's third law: for every action, there is an equal and opposite reaction. Karma is said to determine our destiny, increasing or decreasing during our lifetime, in accordance with the goodness or badness of our thoughts, words, and deeds.[994]

- **Buddhists** believe the human condition is governed by four spiritual laws: (1) suffering is universal, (2) worldly desires cause suffering, (3) suffering is ended by annihilating worldly desires, (4) accomplishing this requires us to follow the Eight-Fold Path. That optimal path is said to consist of right belief, right aims, right speech, right actions, right occupation, right endeavor, right mindfulness, and right meditation.[995]

- **Christians** believe the spiritual world obeys laws decreed by a single, rational deity. This God is not entirely scrutable, but aspects of his intentions and coherence are said to be inferable from (1) carefully preserved biblical records, which include revelations from his prophets and eye-witness testimony from the time of Jesus Christ, his earthly incarnation; and (2) the extraordinary, supernatural-like orderliness of the universe. As Albert Einstein once remarked, ". . . a priori, one should expect a chaotic world, which cannot be grasped by the mind in any way." Therefore, "I consider the comprehensibility of the world . . . as a miracle or as an eternal mystery."[996]

The Pew Research Center reports 84% of all humans on the planet live by religious beliefs such as these.[997] Even among the remaining 16%—persons who are unaffiliated with any specific religion, AKA *nones*—only a tiny fraction are atheists. In the United States, just 13% of nones profess atheism. The rest aren't sure what they believe.[998]

From time immemorial—including the violent, antiestablishment 1960s—secularists have predicted the demise of spiritual thought. But their forecasts have proven to be nothing more than wishful thinking.

Yet, they're still at it.

"Are we naive today to believe that the gods of the present will survive and be here in a hundred years?" asked Dan Brown at the 2017

Frankfurt Book Fair, while promoting his new novel, *Origin*. "Our need for that exterior god, that sits up there and judges us . . . will diminish and eventually disappear."[999]

But it is Brown who is being terribly naive, not to mention woefully ignorant of human nature and human history.

Not only have the world's major belief systems and deities persisted in the face of secular education and technological growth for *thousands* of years, they show no signs of diminishing or disappearing anytime soon—certainly not within a hundred years. In fact, Pew reports the world's nones are the ones in decline. Between now and 2060, their number is projected to drop from 16% to 13%.[1000]

"The secularization thesis of the 1960s—I think that was hopeless," says Roger Trigg, a well-respected Oxford University philosopher.[1001] In 2011 he co-directed the Cognition, Religion and Theology Project, a massive, multimillion-dollar, twenty-nation study of world religions. Its main finding, he reports, is that "attempts to suppress religion are likely to be short-lived, as human thought seems to be rooted to religious concepts, such as the existence of supernatural agents or gods, and the possibility of an afterlife or pre-life."[1002]

Religious thought, therefore, deserves and demands to be taken seriously, especially in today's technologically infatuated world. "Religion is not just something for a peculiar few to do on Sundays instead of playing golf," explains Trigg. "Religion is a common fact of human nature across different societies." That means, "It's got to be reckoned with. You can't just pretend it isn't there."[1003]

Above all, we must reckon with what is arguably the most significant spiritual consensus of all: *life is sacred*. At a minimum, this means believing nothing—not even scientific curiosity or technical know-how—trumps life's unique standing. Nothing other than life merits a higher priority or greater protection from any and all assaults.

"Life is sacred," averred Albert Einstein in his 1949 biography, *The World as I See It*, "that is to say, it is the supreme value, to which all other values are subordinate."[1004]

Life is also the grandest mystery of all. For us to even begin comprehending its origin and fate, we need more than mere logic. We need the help of spiritual thinking to discern universal truths unfathomable to the scientific method.

"Even the weakest and most vulnerable, the sick, the old, the unborn and the poor," declares Pope Francis, "are masterpieces of God's creation, made in his own image, destined to live forever, and deserving of the utmost reverence and respect."[1005]

## THE SCIENTIFIC WORLDVIEW

There are many ways to recount the long and distinguished history of science, which is a subject near and dear to my heart. I have studied and written about it for many decades.

As I see it, science is like a living organism that has matured in stages. *Six stages*, over thousands of years to be precise. Here are my summaries of them.

- **Stage 1**: Ancient peoples swung their eyes skyward and *observed* the heavens. The three magi, whose pilgrimage to Bethlehem is reported in the Christian Gospels, were diligent students of the night sky, because they believed it telegraphed important messages, like a cosmic-sized LED screen. Ancient Mayan astronomer-priests used their intimate knowledge of the sun and moon to predict eclipses, which helped cement their privileged social and religious statuses. Ancient Egyptians used their observational knowledge of the skies to tell farmers when to plant and when to expect flooding from the Nile River.
- **Stage 2**: During the fourth century BC, Aristotle *cataloged* Earth's living organisms, based on their similar traits. For instance, he grouped animals according to whether they had blood or not—then further divided them into those that walked, flew, or swam. Aristotle also

catalogued heavenly bodies, sorting out sun, moon, planets, stars, and comets—placing *Earth* at the center of his cosmology. As we'll see shortly, the legendary Greek philosopher's theories were so persuasive, they survived well into the seventeenth century.[1006]

- **Stage 3**: During the Middle Ages, many natural philosophers sat around *theorizing* and *arguing* about how the universe works. Typical of these armchair philosophers were the scholastics, who taught at universities throughout Europe. The oft-stated accusation that they used scientific-religious reasoning to theorize and argue about how many angels could conceivably dance on the point of a needle is probably not true. But it illustrates how scientists at this early stage used mostly logic to ferret out truths.

- **Stage 4**: Other medieval natural philosophers, not content merely to sit and speculate endlessly about how things work, *experimented* to see which hypotheses were best supported by reliable data. Among those who led the way was Alhazen (short for Abu Ali al-Hasan ibn al-Haytham), a devout Sunni astronomer and mathematician.[1007] His experimental work on light, optics, and vision preceded Isaac Newton's own masterful contributions to those subjects by more than 600 years.

- **Stage 5**: As the Middle Ages gave way to the Renaissance, two Englishmen—Franciscan monk-philosopher Roger Bacon and lawyer-philosopher Francis Bacon (no apparent relation)—*codified* all of science's earlier, formative processes into a standardized scientific method that anyone, anywhere could follow. Galileo Galilei and Isaac Newton tweaked and executed the scientific method to near-perfection with their landmark experiments, laying the foundation for today's classical, experimental physics.[1008]

- **Stage 6**: Throughout its first five stages of growth, science was extremely religion-friendly, dominated by people with a burning desire to understand the cosmos, whose origins and operations they believed to be divine.[1009] These included luminaries such as William of Ockham, Nicolaus Copernicus, and Gregor Mendel, all of whom were both formidable scientists *and* devout clerics.[1010]

But even before the appearance of Darwin's theory of natural selection—according to which we descended from apes, which themselves evolved from shrew-like mammals—science gradually tore loose from its religious roots and became *secularized*. By the twentieth century the divorce was quite final. Scientific thinking, like today's Icelandic culture, was scrubbed of God and religious thinking. Today, science is controlled by a secular philosophy known as rational materialism.[1011]

I, for one, defend science's right to decide what it wishes to be—and not be. Also, please note: science has evicted God and spiritual thinking from its experimental method, but it is *neutral* on the question of God's existence and the veracity of religious thinking. Both matters are well outside of science's jurisdiction to prosecute.

All that notwithstanding, there are people who insist on treating science as a religion—the very thing it strives not to be. Their religion is called *scientism*.

In a nutshell, scientismists believe the universe is made entirely of matter and physical energy. That's it. There's nothing more to it than what is detectable by *human* senses and *human* instruments.

Clearly, as religions go, scientism is exceptionally small-minded. No other world religion envisions a cosmos defined and delimited by the five human senses. Indeed, such a belief smacks of anthropocentricism, which is decidedly unscientific.

Scientism is also self-deflating, insofar as it believes—to paraphrase the late Stephen Hawking—humans are an *unremarkable* species that evolved by chance on an *unremarkable* planet in an *unremarkable* solar system at the outskirts of an *unremarkable* galaxy.[1012] What, then, does that make scientism and its adherents, except quite unremarkable?

The one feature scientism has in common with all other religions is that it requires faith from its adherents. Despite their professed strict allegiance to logic and materialism, despite whatever scientific evidence they cite in favor of their anthropocentric worldview, scientismists cannot prove the universe is made of *only* stuff humans can see, hear, smell, taste, and touch. They must *believe* it is so.

"Scientism," observes Thomas Burnett, a philosopher, science historian, and writer at the National Academy of Sciences, "is a speculative worldview about the ultimate reality of the universe and its meaning."[1013]

Being such a flawed, speculative religion, scientism would normally not be worth discussing. But, alas, it is chiefly responsible for flipping humanity's traditional values on their head.

Today, scientism's religious influence is swaying many people to believe *science* is sacred, above everything else—including life. The result is that in laboratories worldwide, *scientific inquiry* is being defended at all costs. And human life is being treated as science's legitimate plaything—as pliant as children's clay and customizable as a sneaker, sexbot, or social-media homepage.

## HOW I SEE OUR FUTURE

As a physicist, I love science; it brings me unspeakable joy. But I don't worship it, and neither should you. Believing science is sacred is not only unscientific, as I've explained, it easily leads to a fanatical intolerance of anyone who challenges science's authority.

Let me explain.

Many people who believe science is sacred also believe it is infallible. Like a Pope. They think questioning a scientific consensus about *anything* is tantamount to heresy, subject to—if not death by fire—death by shaming. Historically, it is an all-too-familiar scenario.

"Eugenists may remember that not many years ago this program for race regeneration was subjected to the cruel ridicule of *stupidity* and *ignorance*." [Emphasis is mine.]

Margaret Sanger spoke those words against people who dared to oppose eugenics, the muscular scientific consensus of the early twentieth century. Today, identical invectives are being hurled at people who dare to question, for example, the scientific consensus that climate change is being driven entirely or mostly by human activity.

"I think that denying climate change is a crime against humanity," declares Monty Python comedian Eric Idle, not joking. "And they should be held accountable in a World Court."[1014]

Bill Nye—another comedian, but with an undergraduate degree in mechanical engineering—agrees, saying, "this extreme doubt about climate change is affecting my quality of life as a public citizen. So I can see where people are very concerned about this, and they're pursuing criminal investigations . . ."[1015]

"This is treason," says environmentalist Robert F. Kennedy Jr. about people who question science's climate-change consensus. "And we need to start treating them as traitors."[1016]

This is not the place for me to present my own thoughts on the subject—although I will say, they are far more nuanced than the ones above. My point is simply that science, while brilliant, is *fallible*. As I've explained in other published writings, it routinely gets things wrong.[1017]

In fact, an errant scientific consensus was the real culprit in the seventeenth-century Catholic Church's prosecution of Galileo—one of the most cited but grossly misunderstood condemnations of religious thinking. Here's what happened.

The theory placing *Earth* at the center of the universe traces back to Aristotle, as I explained earlier. During the thirteenth century, Thomas Aquinas, an Italian Dominican Friar and a devout Aristotelian, assiduously integrated his beloved Greek philosopher's geocentric cosmology into the warp and woof of official Catholic theology.[1018]

In the sixteenth century, Nicolaus Copernicus—a canon of the Catholic church and an amateur astronomer—defied conventional wisdom by lobbying for a heliocentric theory, which placed the *sun* at the center of everything. Scientists and university scholars of the day heartily debated his arguments, but in the end were not persuaded by them because, for one thing, astronomical observations didn't strongly favor Copernicus's theory.[1019]

At the start of the seventeenth century, therefore, the geocentric theory still ruled the roost. The vast majority of learned people continued defending Aristotle's belief that Earth was at the center of the cosmos.

In 1633, therefore, when the Church condemned Galileo for supporting Copernicus's heliocentric theory, it was defending not just Scripture, but the scientific consensus of the day.[1020] *Far from being the bad guy, orthodox religion was siding with science.*

As a physicist, I completely understand how our contemporary world—which prides itself on progress and looks to science for instruction about *everything*—would fall prey to villainizing God and spiritual thinking, to looking down on the two as quaint and unnecessary relics of the past. Many years ago, as I embarked on my graduate studies in high-energy physics at Cornell University, I thought and behaved the very same way.[1021]

But now I see more clearly: *truth is imperishable.* Panning God and spiritual thinking simply because they are ancient ideas is not reasonable.

Instead, a reasonable person asks questions such as these: Is there any defensible, contemporary evidence in favor of these ideas from our deep past? Do these hoary ideas offer any lofty, philosophical lessons for us today? Do these age-old ideas have any *practical, day-to-day* value for us moderns who must navigate a complicated world of web, robot, spy, and Frankenstein?

The answers, I sincerely believe, are yes, yes, and yes.

Indeed, as we plunge deeper into the twenty-first century, I believe the future will be great or grim not solely because of what we do with our new technological prowess. It will be great or grim, depending on what we do with our ancient biblical wisdom.

Put another way, I believe the livability of our future now rests not on which AI gadgets or jaw-dropping innovations we choose to embrace—or even on successfully colonizing another planet for a fresh start, as many doomsayers are frantically recommending.[1022] It rests on which *beliefs* and *values* we choose to keep or to jettison in the years ahead.

In short, our IQ alone isn't going to save us. If in the years ahead we fail to create a loving, nurturing, meaningful future for everyone; if we cherish only scientific inquiry and not the sanctity of life—*all* life, not just human—I believe our vaunted intelligence might end up destroying us.

"As a species, we've somehow survived large and small ice ages, genetic bottlenecks, plagues, world wars, and all manner of natural disasters," observes American naturalist Diane Ackerman, "but I sometimes wonder if we'll survive our own ingenuity."[1023]

I believe if we fail to create a loving, nurturing, meaningful future for ourselves, it won't be because we weren't *smart* enough; it will be because we weren't *wise* enough. I believe wisdom—such as memorialized in our species' unique, rich library of God-breathed thought—is the only thing that can save us now.

"And what do you benefit if you gain the whole world but lose your own soul?" asked a young Christian named John Mark during the first century AD—way before Facebook, iPhones, Alexa, or Arctic apples.[1024]

If we ever decide to ignore such timeless spiritual wisdom or disparage it as being outdated, I predict we will regret it. One day we will stop, look back, and realize that in our mad dash toward an overhyped, self-centered scientific utopia, we left *ourselves* behind, by abandoning the unique soul and spirit of our species.

# PRACTICAL TIPS

## GREAT FUTURE . . . GRIM FATE?

## IT'S UP TO *YOU!*

*Read, add to, and interact with our constantly updated list of practical tips on how you can help shape a great future and avoid a grim fate.*

Go to www.michaelguillen.com

# BIBLIOGRAPHY

## INTRODUCTION

### GREAT OR GRIM?

1.  http://www.telegraph.co.uk/education/expateducation/11231600/Distance-learning-is-now-open-to-all-thanks-to-the-internet.html
2.  http://www.healthcarebusinesstech.com/how-telemedicine-is-revolutionizing-health-care/
3.  https://www.reuters.com/article/us-facebook-politics/facebook-says-it-cant-guarantee-social-media-is-good-for-democracy-idUSKBN1FB14G
4.  https://nypost.com/2016/12/22/your-social-media-addiction-is-giving-you-depression/
5.  http://www.news.com.au/lifestyle/relationships/dating/as-many-sexual-partners-as-they-can-get-dating-apps-fuelling-rise-in-casual-sex/news-story/7bb94ffbf5ddec69fd5a114aabed54db
6.  https://www.theguardian.com/technology/2017/jan/27/rising-numbers-of-criminals-are-using-facebook-to-document-their-crimes
7.  http://www.miamiherald.com/news/nation-world/national/article129120064.html
8.  https://www.huffingtonpost.com/anna-akbari-phd/why-the-internet-makes-us_b_5325133.html

9.    https://www.theverge.com/2017/12/11/16761016/former-facebook-exec-ripping-apart-society

10.    http://www.electronicdesign.com/communications/internet-things-can-save-50000-lives-year

11.    http://www.bbc.com/news/technology-39901382

12.    https://www.newscientist.com/article/mg23531410-700-in-the-darkening-web-misinformation-is-the-most-powerful-cyber-weapon/?utm_term=Autofeed&utm_campaign=Echobox&utm_medium=Social&cmpid=SOC%7CNSNS%7C2017-Echobox&utm_source=Twitter#link_time=1504347553

13.    http://abcnews.go.com/Health/w_MindBodyNews/paralyzed-woman-moves-robotic-arm-mind/story?id=16353993

14.    https://gizmodo.com/bionic-eyes-can-already-restore-vision-soon-theyll-mak-1669758713

15.    http://www.breitbart.com/news/bionic-pancreas-shows-success-at-controlling-blood-sugar/

16.    https://www.wired.com/2017/05/google-uses-ai-create-1000s-new-musical-instruments/

17.    https://www.bloomberg.com/news/articles/2017-03-29/dominos-will-begin-using-robots-to-deliver-pizzas-in-europe

18.    http://stocknewspress.com/2018/01/05/lg-electronics-to-sell-robots-to-replace-hotel-airport.html

19.    https://www.cebglobal.com/talentdaily/amazons-robot-workforce-has-increased-by-50-percent/

20.    https://www.bloomberg.com/news/articles/2017-11-29/robots-are-coming-for-jobs-of-as-many-as-800-million-worldwide

21.    http://www.mirror.co.uk/news/world-news/robot-doctors-come-step-closer-11542581

22.    https://www.technologyreview.com/the-download/609945/robot-surgeons-are-stealing-training-opportunities-from-young-doctors/

23.    https://www.inc.com/kevin-j-ryan/tim-berners-lee-computers-will-create-new-companies.html?cid=mustread2

24. https://www.cnbc.com/2016/11/04/elon-musk-robots-will-take-your-jobs-government-will-have-to-pay-your-wage.html
25. http://www.bbc.com/news/technology-30290540
26. https://www.axios.com/ges-jeff-immelt-robots-wont-kill-human-jobs-1513301292-bc666afb-ca9e-4bc4-99b4-3406b263bc93.html
27. https://www.forbes.com/sites/johntamny/2015/03/01/why-robots-will-be-the-biggest-job-creators-in-history/#4a04e9262d46
28. https://www.cnbc.com/2017/02/13/elon-musk-humans-merge-machines-cyborg-artificial-intelligence-robots.html
29. http://listverse.com/2013/08/03/10-new-technologies-that-will-usher-in-an-era-of-cyborgs/
30. https://www.pcworld.com/article/3155000/consumer-electronics/hyundais-wearable-robots-could-make-you-superhuman.html
31. https://futurism.com/darpa-is-planning-to-hack-the-human-brain-to-let-us-upload-skills/
32. https://www.nbcnews.com/mach/technology/godlike-homo-deus-could-replace-humans-tech-evolves-n757971
33. https://www.wsj.com/articles/gm-to-test-fleet-of-electric-cars-in-new-york-1508212801
34. https://www.reuters.com/article/us-autos-selfdriving-uber/self-driving-uber-car-kills-arizona-woman-crossing-street-idUSKBN1GV296
35. https://www.bloomberg.com/news/articles/2016-09-22/robot-rides-may-force-error-prone-human-motorists-off-the-road
36. https://www.usatoday.com/story/money/cars/2017/11/23/self-driving-cars-programmed-decide-who-dies-crash/891493001/
37. https://www.wsj.com/articles/supersmart-robots-will-outnumber-humans-within-30-years-says-softbank-ceo-1488193423
38. https://www.theregister.co.uk/2017/01/13/eu_treat_robots_as_people/
39. http://news.rice.edu/2016/02/14/when-machines-can-do-any-job-what-will-humans-do/

40.    https://www.komando.com/happening-now/410905/amazons-alexa-helps-catch-thief-red-handed

41.    http://www.cnn.com/2017/01/12/tech/voice-technology-internet-of-things-privacy/index.html

42.    https://nakedsecurity.sophos.com/2013/08/13/london-says-media-companys-spying-rubbish-bins-stink/

43.    https://www.digitaltrends.com/mobile/is-your-smartphone-listening-to-your-conversations/

44.    https://www.engadget.com/2017/07/12/amazon-developers-private-alexa-transcripts/

45.    https://www.reuters.com/article/us-apple-iphone-privacy-analysis/app-developer-access-to-iphone-x-face-data-spooks-some-privacy-experts-idUSKBN1D20DZ

46.    https://thehackernews.com/2017/05/ultrasonic-tracking-signals-apps.html

47.    https://motherboard.vice.com/en_us/article/qkm48b/how-this-internet-of-things-teddy-bear-can-be-remotely-turned-into-a-spy-device

48.    https://arstechnica.com/tech-policy/2017/11/an-alarming-number-of-sites-employ-privacy-invading-session-replay-scripts/

49.    https://www.theguardian.com/technology/2017/jul/03/facebook-track-browsing-history-california-lawsuit

50.    http://www.news.com.au/technology/online/security/creepy-new-website-makes-its-monitoring-of-your-online-behaviour-visible/news-story/3e60aa78393affd8e89a6692dbbc426f

51.    https://www.cbsnews.com/news/casper-mattress-startup-sued-for-wiretapping-site-visitors/

52.    http://www.nkytribune.com/2017/04/keven-moore-surveillance-cameras-are-everywhere-providing-protection-but-not-much-privacy/

53.    https://www.hackread.com/website-streams-from-private-security-cameras/

54.  https://www.wsj.com/articles/the-all-seeing-surveillance-state-feared-in-the-west-is-a-reality-in-china-1498493020

55.  http://www.ibtimes.co.uk/britain-cctv-camera-surveillance-watch-london-big-312382

56.  https://ipvm.com/reports/america-cctv-recording

57.  http://www.cnn.com/2013/04/26/tech/innovation/security-cameras-boston-bombings/index.html

58.  http://www.vocativ.com/402771/ai-body-cams-cops-google/

59.  http://money.cnn.com/video/technology/security/2013/08/01/t-tv-is-watching-you.cnnmoney/index.html

60.  http://www.businessinsider.com/walmart-is-developing-a-robot-that-identifies-unhappy-shoppers-2017-7

61.  http://www.nydailynews.com/news/national/amazon-driver-caught-video-pooping-front-home-article-1.3670807

62.  https://www.fastcompany.com/3069264/congress-fbi-face-recognition-real-time-street-lineup

63.  https://arstechnica.com/tech-policy/2016/06/smile-youre-in-the-fbi-face-recognition-database/

64.  https://www.washingtonpost.com/local/public-safety/the-new-way-police-are-surveilling-you-calculating-your-threat-score/2016/01/10/e42bccac-8e15-11e5-baf4-bdf37355da0c_story.html?utm_term=.87fe877ea140

65.  https://www.thesun.co.uk/tech/3307230/u-s-navy-funds-development-of-robot-surveillance-system-which-can-spy-on-humans-in-terrifying-detail/

66.  http://mysteriousuniverse.org/2017/01/cyborg-dragonfly-developed-for-spying/

67.  http://www.breitbart.com/video/2017/10/10/talking-drone-trying-to-lure-kids-from-ohio-playground/

68.  http://gephardtdaily.com/local/drone-flying-peeping-tom-bedroom-windows/

69.  http://neurosciencenews.com/machine-learning-thought-6974/

70. http://www.dailymail.co.uk/sciencetech/article-5197747/AI-detects-expressions-tell-people-lie-court.html

71. https://www.theverge.com/2017/10/30/16570148/neuroscience-suicide-mental-health-brain-imagining-fmri-machine-learning-algorithm

72. http://www.cnn.com/2017/03/08/politics/james-comey-privacy-cybersecurity/

73. http://boston.cbslocal.com/2017/10/24/brigham-and-womens-dna-trial-babyseq/

74. http://metro.co.uk/2017/11/28/worlds-first-spy-condom-collects-intimate-data-during-sex-and-tells-men-whether-their-performance-is-red-hot-or-a-total-flop-7116049/

75. https://www.thesun.co.uk/news/2626083/bae-systems-turns-earths-atmosphere-into-massive-surveillance-system-using-lasers/

76. http://www.telegraph.co.uk/science/2016/11/09/paralysed-people-could-walk-again-instantly-after-scientists-pro/

77. http://www.telegraph.co.uk/science/2017/08/04/tiny-human-brains-grown-lab-could-one-day-used-repair-alzheimers/

78. http://www.foxnews.com/science/2017/11/08/scientists-implant-tiny-human-brains-into-rats-spark-ethical-debate.html

79. https://www.usatoday.com/story/news/world/2017/11/17/italian-doctor-says-worlds-first-human-head-transplant-imminent/847288001/

80. https://www.irishtimes.com/news/genetic-engineering-way-forward-for-medical-science-or-sinister-threat-to-all-our-futures-1.46757

81. https://www.newscientist.com/article/2119252-gene-editing-has-saved-the-lives-of-two-children-with-leukaemia/

82. http://www.sciencemag.org/news/2017/11/boy-rare-disease-gets-new-skin-thanks-gene-corrected-stem-cells

83. https://www.wsj.com/articles/the-gene-editors-are-only-getting-started-1499461756

84. https://www.technologyreview.com/s/600774/top-us-intelligence-official-calls-gene-editing-a-wmd-threat/

85. https://www.newscientist.com/article/dn27402-first-human-embryos-genetically-modified-more-will-come/

86. http://www.foxnews.com/science/2017/03/16/chinese-researchers-announce-designer-baby-breakthrough.html

87. https://www.technologyreview.com/s/535661/engineering-the-perfect-baby/

88. https://www.newscientist.com/article/2107219-exclusive-worlds-first-baby-born-with-new-3-parent-technique/

89. https://www.theverge.com/2014/3/27/5553044/first-functional-eukaryotic-chromosome

90. http://www.newsweek.com/2017/07/07/natural-selection-new-forms-life-scientists-altering-dna-629771.html

91. https://www.marketwatch.com/story/how-upgrading-humans-will-become-the-next-billion-dollar-industry-2017-04-03

92. http://www.telegraph.co.uk/science/2016/12/15/scientists-reverse-ageing-mammals-predict-human-trials-within/

93. https://www.nbcnews.com/mach/science/death-may-not-be-so-final-thanks-these-creepy-technologies-ncna821596

94. https://news.nationalgeographic.com/2017/01/human-pig-hybrid-embryo-chimera-organs-health-science/

95. http://www.iflscience.com/health-and-medicine/human-animal-hybrids-investigated-source-transplant-organs/

96. http://www.iflscience.com/health-and-medicine/human-ear-has-been-grown-rat/

97. https://www.dallasnews.com/news/zika-virus/2017/07/14/genetically-engineered-mosquitoes-wipe-zika-dallas-county-oppose-local-trials

98. https://www.nytimes.com/2017/03/20/science/revive-restore-extinct-species-dna-mammoth-passenger-pigeon.html?rref=collection%2Fsectioncollection%2Fscience&action=click&contentCollection=science&region=rank&module=package&version=highlights&contentPlacement=1&pgtype=sectionfront&_r=0&mtrref=undefined

99.    http://www.realclearlife.com/science/peter-thiel-funding-effort-bring-woolly-mammoths-back-life/

100.    http://pittsburgh.cbslocal.com/2017/11/13/texas-company-cloning-pets/

101.    https://www.wsj.com/articles/monsanto-bets-on-next-phase-of-high-tech-crops-but-its-not-alone-1494162000

102.    https://www.cbsnews.com/news/what-have-they-done-to-our-food-28-02-2001/

103.    http://www.sciencemag.org/news/2016/05/once-again-us-expert-panel-says-genetically-engineered-crops-are-safe-eat

104.    https://www.wired.com/story/genetically-modified-arctic-apple-targets-consumers-not-farmers/

105.    https://www.washingtonpost.com/news/morning-mix/wp/2017/03/27/scientists-convert-spinach-leaves-into-human-heart-tissue-that-beats/?utm_term=.bec9f04fac69

106.    Story appeared in the *Chicago Tribune* on Sunday, September 25, 1910. See also: https://books.google.com/books?id=1MlOAAAAYAAJ&pg=PA461&lpg=PA461&dq=Dr+C+G+Davis+Chicago+Christian+Endeavor+Society&source=bl&ots=5qHUAdDxJN&sig=b4SwLSqOk4MH1DF3OKAG5eJf6jg&hl=en&sa=X&ved=0ahUKEwj7uKew6sXYAhVBw4MKHcQtAqkQ6AEIMTAC#v=onepage&q=Dr%20C%20G%20Davis%20Chicago%20Christian%20Endeavor%20Society&f=false, p 461.

107.    https://www.theatlantic.com/health/archive/2013/03/how-we-realized-putting-radium-in-everything-was-not-the-answer/273780/ and http://blog.nyhistory.org/get-me-a-radium-highball-new-york-and-the-radium-craze/

108.    https://www.scientificamerican.com/slideshow/radioactive-products/

109.    Davis, Charles Gilbert. (1921). "Radium and Its Therapeutic Application." *American Journal of Clinical Medicine* 85–92 and Davis, Charles Gilbert. (1922). "Hot Radium Springs, With Special Reference to the Waunita Springs in Colorado." *American Journal of Clinical Medicine* 491–496.

110.  Many references. See, for example: http://www.waterburyobserver
      .org/wod7/node/2723

111.  White, E. B. (1977). "'Coon Tree'" (14 June 1956), The Points of
      My Compass: Letters from the East, the West, the North, the
      South (1962)" In *Essays of E.B. White* and https://libquotes
      .com/e-b-white/quote/lbb8v8u

112.  https://www.nytimes.com/2014/03/16/science/billionaires-with-
      big-ideas-are-privatizing-american-science.html

113.  https://www.goodreads.com/work/quotes/43314655-you-re-
      never-weird-on-the-internet-almost

# WEB

## MEMORY LANE

114.  http://www.guinnessworldrecords.com/world-records/farthest-
      distance-travelled-by-a-human-voice

115.  https://marl.smusic.nyu.edu/papers/boren_dissertation_nyu.pdf

116.  http://americanhistory.si.edu/collections/search/object/
      nmah_713485

117.  *Illinois Bell Magazine: Volumes 3-4*, p 34. https://books.google.com/
      books?id=oGo2AQAAMAAJ&pg=RA2-PA34&lpg=RA2-PA3
      4&dq=%22Ahoy!+Ahoy!+Mr.+Watson,+are+you+there?+Do+yo
      u+hear+me?%22&source=bl&ots=UvY2Jd21QY&sig=N3aSGw
      T4TeQmL3sG5mIR21awIGw&hl=en&sa=X&ved=0ahUKEwi
      kjO2soanSAhUFWCYKHUhbDcsQ6AEIMTAF#v=onepage&
      q=%22Ahoy!%20Ahoy!%20Mr.%20Watson%2C%20are%20
      you%20there%3F%20Do%20you%20hear%20
      me%3F%22&f=false. See also: http://www.nytimes.com/learning/
      general/onthisday/big/0125.html

118.  See, for example, http://www.telcomhistory.org/vm/
      scienceLongDistance.shtml and https://hackaday.com/2016/03/18/
      what-lies-beneath-the-first-transatlantic-communications-cables/

119.  https://www.wired.com/1996/12/ffglass/

120. http://www.sparkmuseum.com/BOOK_HERTZ.HTM
121. http://www.history.com/this-day-in-history/marconi-sends-first-atlantic-wireless-transmission
122. http://www.heritage.nf.ca/articles/society/marconi-guglielmo.php
123. http://www.hammondmuseumofradio.org/fessenden-2006-recreation.html
124. http://theinventors.org/library/inventors/bl_television_timeline.htm and https://www.facebook.com/HooverPresLib/videos/116047015077905/
125. http://www.computerhistory.org/babbage/history/, http://www.charlesbabbage.net/, and https://www.youtube.com/watch?v=BlbQsKpq3Ak (remarkable video of a replica of Babbage's mechanical computer in action)
126. http://www.computerhistory.org/timeline/1942/
127. http://www.computerhistory.org/revolution/birth-of-the-computer/4/78/323
128. http://www.thocp.net/hardware/eniac.htm
129. http://www.popularmechanics.com/space/rockets/a24429/hidden-figures-real-story-nasa-women-computers/
130. https://history.nasa.gov/sputnik/
131. http://www.vanityfair.com/news/2008/07/internet200807?currentPage=3
132. http://internethalloffame.org/inductees/jcr-licklider
133. http://web.stanford.edu/dept/SUL/library/extra4/sloan/mousesite/Secondary/Licklider.pdf
134. http://newsroom.ucla.edu/releases/birthplace-of-the-internet-celebrates-111333
135. http://www.computerhistory.org/timeline/1976/, https://www.wired.com/2012/12/queen-and-the-internet/, and https://www.amazon.com/gp/search?index=books&linkCode=qs&keywords=9781606499931
136. https://www.w3.org/History/1989/proposal.html
137. http://internethalloffame.org/blog/2012/06/06/berners-lee-world-finally-realizes-web-belongs-no-one

138.  https://www.amazon.co.uk/Weaving-Web-Present-Future-Inventor/dp/0752820907

139.  https://www.w3.org/People/Berners-Lee/FAQ.html#Spelling

140.  Berners-Lee prefers the term "URI," Uniform Resource Identifier. For a discussion of that and why he chose to begin Web addresses with "http://" see endnote 19.

141.  https://www.w3.org/People/Berners-Lee/FAQ.html

142.  http://science.sciencemag.org/content/early/2011/02/09/science.1200970

143.  https://hackercombat.com/the-best-10-deep-web-search-engines-of-2017/

144.  http://www.pcadvisor.co.uk/how-to/internet/what-is-dark-web-how-access-dark-web-deep-3593569/ and https://www.quora.com/What-are-the-other-browsers-to-access-the-deep-dark-web-other-than-Tor

145.  https://www.wired.com/2012/06/sir-tim-berners-lee/

## STAR POWER

146.  I say "more" because if the 100 people you add per day each reaches out to another 100 people per day, then after three days your total audience will be $(100 \times 100 \times 100) + (100 \times 100 \times 100) + (100 \times 100 \times 100)$, i.e., 3,000,000 people.

147.  http://www.cl.cam.ac.uk/coffee/qsf/coffee.html

148.  http://www.bbc.com/news/technology-20439301

149.  https://www.thevintagenews.com/2016/08/26/priority-trojan-room-coffee-pot-inspiration-worlds-first-webcam/

150.  https://www.youtube.com/watch?v=zGxwbhkDjZM and http://www.businessinsider.com/sweet-brown-apple-lawsuit-2013-3

151.  https://www.youtube.com/watch?v=bFEoMO0pc7k

152.  https://www.instagram.com/p/_AcyN5Avvt/?hl=en

153.  http://www.unomodels.com/model/1864-cindy-kimberly/

154.  https://www.instagram.com/wolfiecindy/?hl=en

155. http://www.dailymail.co.uk/news/article-3440232/Spanish-17-year-old-famous-Justin-Bieber-posted-photograph-saying-OMG-quits-3-hour-babysitting-job-signed-model-agency.html
156. https://www.youtube.com/watch?v=9bZkp7q19f0
157. https://www.theguardian.com/world/2012/oct/24/psy-gangnam-style-united-nations
158. http://bigstory.ap.org/article/3a9dc181c2924fd49136521c8beb9af4/phones-and-social-media-turn-consumers-whistleblowers
159. http://foreignpolicy.com/2017/04/11/chinese-blame-america-for-united-airlines/
160. http://www.latimes.com/business/la-fg-china-united-backlash-20170411-story.html
161. https://www.nytimes.com/2017/04/27/business/united-david-dao-settlement.html
162. https://www.biography.com/people/jeff-bezos-9542209
163. https://suebrewton.com/2016/06/30/no-john-d-rockefeller-did-not-write-that/
164. http://mashable.com/2011/07/22/facts-amazon-com/#CozqZCUj7qqH
165. http://mashable.com/2011/07/22/facts-amazon-com/#CozqZCUj7qqH
166. http://www.investopedia.com/university/jeff-bezos-biography/jeff-bezos-success-story.asp
167. https://www.washingtonpost.com/lifestyle/style/jeffrey-bezos-washington-posts-next-owner-aims-for-a-new-golden-era-at-the-newspaper/2013/09/02/30c00b60-13f6-11e3-b182-1b3bb2eb474c_story.html?utm_term=.8d2813fa9f50
168. http://www.macrotrends.net/stocks/charts/AMZN/market-cap/amazon-inc-market-cap-history
169. https://www.recode.net/2018/2/21/17035706/amazon-jeff-bezos-worth-walmart-stock-market-value-growth
170. http://fortune.com/2018/02/15/amazon-microsoft-third-most-valuable-company/

171. http://money.cnn.com/2018/01/09/technology/jeff-bezos-richest/index.html

172. http://time.com/money/5334441/amazon-jeff-bezos-net-worth-prime-day-2018/

173. https://www.seattletimes.com/business/amazon/amazon-pulled-into-another-sales-tax-fight-as-states-go-after-third-party-sellers/ and https://www.statista.com/statistics/259782/third-party-seller-share-of-amazon-platform/

174. http://fortune.com/2017/10/26/a-record-amount-of-brick-and-mortar-stores-will-close-in-2017/

175. https://finance.yahoo.com/news/staggering-amount-u-s-retail-stores-closed-2017-161401876.html

176. http://www.businessinsider.com/stores-closing-in-2018-2017-12

177. https://www.bloomberg.com/news/articles/2017-04-07/stores-are-closing-at-a-record-pace-as-amazon-chews-up-retailers

178. http://fortune.com/2017/04/28/5-reasons-amazon-physical-stores/

179. https://www.wsj.com/articles/the-mall-of-the-future-will-have-no-stores-1497268801

180. https://www.gofundme.com/success

181. https://www.gofundme.com/fromcompton2harvard

182. https://www.kickstarter.com/projects/smithsonian/reboot-the-suit-bring-back-neil-armstrongs-spacesu/?utm_source=smithmag

## THE WILD WILD WEB

183. http://time.com/4457110/internet-trolls/

184. http://time.com/4457110/internet-trolls/

185. www.todayinsci.com

186. http://www.huffingtonpost.com/anna-akbari-phd/why-the-internet-makes-us_b_5325133.html

187. http://www.legendsofamerica.com/ks-boothill.html

188.  http://latimesblogs.latimes.com/technology/2011/11/although-69-of-teenagers-who-use-social-networking-websites-say-their-peers-are-mostly-kind-to-one-another-online-88-said.html

189.  https://www.amazon.com/exec/obidos/ASIN/1416575987/understandi0d-20

190.  https://lendedu.com/blog/millennials-instagram-narcissistic-social-media-platform/

191.  https://lendedu.com/blog/millennials-instagram-narcissistic-social-media-platform/

192.  https://lendedu.com/blog/accurately-social-media-portray-life-millennials/

193.  https://www.technologyreview.com/s/602862/data-scientists-chart-the-tragic-rise-of-selfie-deaths/

194.  http://www.fox26houston.com/news/teen-killed-seconds-after-unbuckling-to-take-a-selfie

195.  https://www.washingtonpost.com/news/worldviews/wp/2016/01/14/more-people-die-taking-selfies-in-india-than-anywhere-else-in-the-world/?utm_term=.9d276ac52948

196.  http://www.dailypioneer.com/columnists/edit/the-deadly-selfie-craze.html

197.  https://www.reuters.com/article/us-life-selfies/selfie-madness-too-many-dying-to-get-the-picture-idUSKCN0R305L20150903

198.  http://quicksprout.wpengine.netdna-cdn.com/wp-content/uploads/2015/03/Ultimate-Guide-to-Creating-Visually-Appealing-Content.jpg

199.  https://animoto.com/blog/business/video-marketing-cheat-sheet-infographic/

200.  http://news.mit.edu/2014/in-the-blink-of-an-eye-0116

201.  http://bionumbers.hms.harvard.edu/bionumber.aspx?id=100706&ver=0

202.  http://en.radiovaticana.va/news/2017/03/21/our_lady_at_the_heart_of_2017_wyd_message_/1300061

203. http://www.cisco.com/c/en/us/solutions/collateral/service-provider/visual-networking-index-vni/complete-white-paper-c11-481360.html

204. http://georgedpcarlin.blogspot.com/2012/04/george-carlin-on-war.html

205. http://georgedpcarlin.blogspot.com/2012/04/george-carlin-on-war.html

206. http://business.time.com/2012/02/01/read-facebook-ceo-mark-zuckerbergs-ipo-letter/

207. https://www.usatoday.com/story/tech/news/2017/04/12/facebook-messenger-hits-new-milestone-12-billion-users/100349968/

208. http://www.stuff.co.nz/technology/social-networking/91729765/social-networks-are-dying-as-messenger-apps-rise-up

209. https://www.amazon.com/gp/search?index=books&linkCode=qs&keywords=9783642215216

210. https://www.theregister.co.uk/2017/03/01/aws_s3_outage/

211. https://www.srgresearch.com/articles/amazon-dominates-public-iaas-paas-ibm-leads-managed-private-cloud

212. https://www.similartech.com/technologies/amazon-s3

213. https://aws.amazon.com/message/41926/

214. http://www.geekwire.com/2017/amazon-explains-massive-aws-outage-says-employee-error-took-servers-offline-promises-changes/

215. https://www.youtube.com/watch?v=7xX_KaStFT8&feature=youtu.be

216. http://www.mapcon.com/timeline-of-computer-viruses

217. http://www.mapcon.com/timeline-of-computer-viruses

218. https://www.alienvault.com/blogs/security-essentials/common-types-of-malware-2016-update

219. https://www.computerworld.com/article/3187520/security/new-ransomware-demanded-high-score-on-anime-style-shooter-game-not-bitcoins.html

220.  https://www.forbes.com/sites/stevemorgan/2015/12/20/
      cybersecurity%E2%80%8B-%E2%80%8Bmarket-reaches-75-
      billion-in-2015%E2%80%8B%E2%80%8B-%E2%80%8
      Bexpected-to-reach-170-billion-by-2020/#701a8c3310c3

221.  http://www.nbcnews.com/tech/tech-news/popular-antivirus-
      program-mistakenly-ids-windows-threat-creating-chaos-n750521

222.  http://nypost.com/1999/11/25/donahue-wants-to-show-a-live-
      killing/

223.  http://www.nydailynews.com/archives/news/60-minutes-airs-dr-
      death-reel-patient-injected-dies-national-tv-article-1.811236

224.  https://www.facebook.com/zuck/posts/10102764095821611

225.  http://www.crimeonline.com/2017/04/17/facebook-killer-strikes-
      as-violence-suicides-plague-social-media-livestreams/

226.  https://www.nytimes.com/2016/05/12/world/europe/periscope-
      suicide-france.html?_r=1

227.  https://www.periscope.tv/about

228.  https://www.usatoday.com/story/news/2017/03/21/chicago-gang-
      rape-teen-streamed-facebook-live/99447884/

229.  http://wp.production.patheos.com/blogs/warrenthrockmorton/
      files/2014/08/FormalCharges-Driscoll-814.pdf

230.  http://blog.seattlepi.com/seattlepolitics/2014/07/29/mars-hill-
      dissenters-plan-protest-release-pussified-nation-driscoll-rant/

231.  http://www.christianitytoday.com/gleanings/2014/august/mark-
      driscoll-crude-comments-william-wallace-mars-hill.html

232.  https://www.theatlantic.com/national/archive/2014/11/houston-
      mark-driscoll-megachurch-meltdown/382487/

233.  https://www.scribd.com/document/265348695/Microsoft-
      Attention-Spans-Research-Report

234.  http://time.com/4457110/internet-trolls/

235.  https://www.theatlantic.com/politics/archive/2017/05/call-out-
      culture-is-stressing-out-college-students/524679/

236.  http://time.com/4457110/internet-trolls/

237. https://newsroom.fb.com/news/2018/01/effect-social-media-democracy/

238. https://newsroom.fb.com/news/2018/01/effect-social-media-democracy/

239. https://www.rsph.org.uk/uploads/assets/uploaded/62be270a-a55f-4719-ad668c2ec7a74c2a.pdf

240. https://www.ncbi.nlm.nih.gov/pmc/articles/PMC5369147/

241. http://www.hmc.org.uk/blog/young-people-rebelling-social-media-survey-reveals/

242. http://www.blog.theteamw.com/2009/11/07/100-things-you-should-know-about-people-8-dopamine-makes-us-addicted-to-seeking-information/

243. https://www.fastcompany.com/1659062/social-networking-affects-brains-falling-love

244. https://www.youtube.com/watch?v=hVDN2mjJpb8

245. http://vt.co/sci-tech/innovation/sean-parker-attacks-mark-zuckerberg-admits-created-monster/

246. https://www.theverge.com/2017/11/9/16627724/sean-parker-facebook-childrens-brains-feedback-loop

247. https://www.theatlantic.com/magazine/archive/2017/09/has-the-smartphone-destroyed-a-generation/534198/

248. https://www.theatlantic.com/magazine/archive/2017/09/has-the-smartphone-destroyed-a-generation/534198/

249. https://www.theatlantic.com/magazine/archive/2017/09/has-the-smartphone-destroyed-a-generation/534198/

250. https://www.youtube.com/watch?v=PMotykw0SIk&feature=youtu.be&t=21m21s and https://www.theverge.com/2017/12/11/16761016/former-facebook-exec-ripping-apart-society

251. https://www.theguardian.com/technology/2017/oct/05/smartphone-addiction-silicon-valley-dystopia

252. https://www.theguardian.com/technology/2017/oct/05/smartphone-addiction-silicon-valley-dystopia

## DISRUPTION AND DECEPTION

253.   https://www.wired.com/2011/02/egypts-revolutionary-fire/
254.   https://shorensteincenter.org/news-coverage-2016-general-election/
255.   https://www.usatoday.com/story/news/world/2016/11/10/world-reaction-trump-papers/93594464/
256.   https://www.wired.com/2011/02/the-internet-explodes-as-egypts-dictator-finally-quits/
257.   http://www.miamiherald.com/news/politics-government/election/article113854758.html
258.   http://www.europarl.europa.eu/stoa/webdav/site/cms/shared/2_events/workshops/2016/20160531/Luc%20Soete.pdf
259.   https://www.prb.org/wp-content/uploads/2015/01/2013-population-data-sheet_eng.pdf
260.   http://www.pewresearch.org/fact-tank/2017/04/06/public-confidence-in-scientists-has-remained-stable-for-decades/
261.   http://www.pewinternet.org/2015/01/29/public-and-scientists-views-on-science-and-society/
262.   http://www.pewinternet.org/2016/10/04/public-views-on-climate-change-and-climate-scientists/
263.   https://www.epw.senate.gov/public/_cache/files/bba2ebce-6d03-48e4-b83c-44fe321a34fa/consensusbusterscompletedocument.pdf
264.   https://www.epw.senate.gov/public/_cache/files/bba2ebce-6d03-48e4-b83c-44fe321a34fa/consensusbusterscompletedocument.pdf
265.   https://www.ewtn.com/library/councils/v1.htm  See Chapter 4, Section 9.
266.   http://www.reformationhappens.com/works/tabletalk/ See part XLIV.
267.   https://www.npr.org/2016/11/20/502437123/how-technology-helped-martin-luther-change-christianity

268. https://www.museeprotestant.org/en/notice/the-reformation-and-the-bible-sola-scriptura/

269. https://www.amazon.com/Structure-Scientific-Revolutions-50th-Anniversary/dp/0226458121

270. http://www.goodreads.com/quotes/51573-no-amount-of-experimentation-can-ever-prove-me-right-a

271. https://www.wsj.com/articles/a-red-team-exercise-would-strengthen-climate-science-1492728579

272. http://webfoundation.org/2017/03/web-turns-28-letter/

273. https://twitter.com/TheAPJournalist/status/8530751174
41769477/photo/1?ref_src=twsrc%5Etfw&ref_url=http%
3A%2F%2Fwww.mediaite.com%2Fonline%2Fbogus-headlines-on-bloomberg-and-chinese-media-risks-starting-wwiii%2F

274. http://www.mediaite.com/online/bogus-headlines-on-bloomberg-and-chinese-media-risks-starting-wwiii/

275. https://themedium.blogs.nytimes.com/2006/07/28/a-single-camera-dramedy-just-in-time-for-the-fall-season/

276. http://www.lg15.com/lgpedia/index.php?title=The_Tolstoy_Principle_(and_Dad_%22talks%22_to_Daniel)

277. http://articles.latimes.com/2006/sep/09/entertainment/et-lonelygirl9

278. https://www.theguardian.com/technology/2016/jun/16/lonelygirl15-bree-video-blog-youtube

279. http://www.nbcnews.com/id/15196982/ns/business-us_business/t/google-buys-youtube-billion/#.WQeLbfnyuM8

280. https://archive.fo/lmHPC

281. https://archive.fo/ByTKo#selection-1737.732-1737.869

282. http://www.huffingtonpost.co.za/verashni-pillay/verashni-pillay-this-is-why-im-sorry_a_22046350/

283. https://archive.fo/ByTKo#selection-1681.0-1681.76

284. https://www.nytimes.com/2017/02/15/arts/fake-news-a-cure-for-wellness-movie.html?rref=collection%2Fbyline%2Fliam-stack&action=click&contentCollection=undefined&region=stream&m

odule=stream_unit&version=latest&contentPlacement=2&pgty pe=collection

285.   https://www.buzzfeed.com/craigsilverman/a-hollywood -film-is-funding-fake-news?utm_term=.ceBzPwnJqe# .mipmP4nD5j

286.   http://archive.is/OxyG1

287.   https://www.nytimes.com/2017/02/16/business/20th-century-fox-fake-news-ad-campaign.html?_r=0&mtrref=undefined

288.   http://www.snopes.com/war-of-the-worlds/

289.   https://www.amazon.com/H-G-Wells-War-Worlds/dp/ 1535001887/ref=pd_lpo_sbs_14_t_0?_encoding=UTF8&psc=1 &refRID=EDDBK627EAPK34SHXF2M

290.   https://www.nytimes.com/2017/04/04/world/middleeast/syria-gas-attack.html?_r=0

291.   https://www.bloomberg.com/news/articles/2017-04-14/russia-says-evidence-growing-syria-chemical-attack-was-staged

292.   http://www.cnn.com/2017/04/13/middleeast/syria-bashar-assad-interview/

293.   http://www.foxnews.com/world/2017/04/13/syrian-first-lady-asma-assads-social-media-posts-reflect-alternate-reality-critics-say.html

294.   https://blog.google/products/search/fact-check-now-available-google-search-and-news-around-world/

295.   https://mediabiasfactcheck.com/mediaite/

296.   https://www.mediaite.com/online/why-does-googles-new-fact-check-feature-seem-to-be-targeting-conservative-sites/

297.   https://www.mediaite.com/online/why-does-googles-new-fact-check-feature-seem-to-be-targeting-conservative-sites/ and http:// www.josephwulfsohn.com/

298.   https://factcheckingday.com/articles/25/april-2-marks-the-second-annual-international-fact-checking-day

299.   https://www.monmouth.edu/polling-institute/reports/ monmouthpoll_us_040218/

300.   https://www.wired.com/story/how-to-watch-mark-zuckerberg-testify-before-congress/

301.   https://www.newsbusters.org/blogs/nb/nb-staff/2018/03/27/
       media-research-center-announces-fact-checking-fact-checkers-
       project
302.   https://leginfo.legislature.ca.gov/faces/billNavClient.xhtml?bill_
       id=201720180AB1104
303.   https://leginfo.legislature.ca.gov/faces/billTextClient.xhtml?bill_
       id=201720180AB1104
304.   http://www.salmanspiritual.com/akbar.cfm
305.   https://townhall.com/tipsheet/mattvespa/2017/04/18/political-
       correctness-ap-opts-not-to-say-that-fresno-shooter-yelled-allahu-
       akbar-writes-english-translation-instead-n2314996
306.   https://www.nytimes.com/2017/04/19/sports/-new-england-
       patriots-visit-white-house-donald-trump.html?smid=t
       w-nytsports&smtyp=cur
307.   https://twitter.com/patriots/status/854858415012806656?lang
       =en
308.   https://www.washingtonpost.com/news/early-lead/wp/2017/
       04/20/patriots-fire-back-at-new-york-times-over-photo-
       comparison-of-trump-obama-white-house-trips/?utm
       _term=.973a252d163d
309.   https://newsroom.fb.com/news/2018/01/effect-social-media-
       democracy/
310.   https://medium.com/facebook-design/designing-against-
       misinformation-e5846b3aa1e2
311.   https://newsroom.fb.com/news/2017/12/news-feed-fyi-updates-
       in-our-fight-against-misinformation/
312.   https://newsroom.fb.com/news/2017/12/news-feed-fyi-updates-
       in-our-fight-against-misinformation/
313.   https://www.facebook.com/zuck/posts/10104380170714571
314.   http://www.kansashistory.us/fordco/lawmen.html
315.   http://webfoundation.org/2017/03/web-turns-28-letter/
316.   http://www.internetworldstats.com/stats.htm

## ROBOT

### MEMORY LANE

317. https://www.forbes.com/sites/bernardmarr/2016/04/05/why-everyone-must-get-ready-for-4th-industrial-revolution/#4ae915613f90

318. https://www.weforum.org/agenda/2016/01/what-is-the-fourth-industrial-revolution/

319. https://www.amazon.com/Fourth-Industrial-Revolution-Klaus-Schwab/dp/1944835008

320. http://www.notable-quotes.com/g/grupen_rod.html

321. https://books.google.com/books?id=GUKbAAAAQBAJ&pg=PT76&lpg=PT76&dq=descartes+%22from+the+mere+arrangement+of+the+machine%27s+organs%22&source=bl&ots=rWB_SPbV6y&sig=qg4pYDS1jXKUPpwdxCPbpHmaX0g&hl=en&sa=X&ved=0ahUKEwif75zO9OXTAhXMQCYKHXviDuMQ6AEINDAD#v=onepage&q=descartes%20%22from%20the%20mere%20arrangement%20of%20the%20machine's%20organs%22&f=false

322. http://www.mahn.ch/collections-arts-appliques-automates

323. https://www.youtube.com/watch?v=IeTOqDb-86s

324. http://news.stanford.edu/news/2001/october24/riskinprofile-1024.html

325. http://www.blackbird.vcu.edu/v1n1/nonfiction/king_e/prayer_introduction.htm

326. http://io9.gizmodo.com/5956937/this-450-year-old-clockwork-monk-is-fully-operational

327. About the outcome of the first five days of creation, the Bible reports: "God saw that it was good." But at the end of the sixth day, after creating Adam and Eve, God reportedly says "it was very good." See https://www.biblegateway.com/passage/?search=Genesis+1&version=NIV

328. http://www.goodreads.com/quotes/558084-imitation-is-the-sincerest-form-of-flattery-that-mediocrity-can

329.  https://ebooks.adelaide.edu.au/c/capek/karel/rur/

330.  http://www.sciencefriday.com/segments/science-diction-the-origin-of-the-word-robot/

331.  His creators always referred to Shakey as "him," not "it." See https://www.sri.com/work/timeline-innovation/timeline.php?timeline=computing-digital#!&innovation=shakey-the-robot

332.  http://www.robothalloffame.org/inductees/04inductees/shakey.html

333.  https://www.sri.com/newsroom/press-releases/sri-internationals-shakey-robot-be-honored-ieee-milestone-computer-history

334.  http://www.computerhistory.org/revolution/artificial-intelligence-robotics/13/289/1229

335.  http://newatlas.com/shakey-robot-sri-fiftieth-anniversary/37668/

336.  http://www.hansonrobotics.com/about/innovations-technology/

337.  https://www.cbsnews.com/news/60-minutes-charlie-rose-interviews-a-robot-sophia/ and https://www.nbc.com/the-tonight-show/video/tonight-showbotics-snakebot-sophia-emotion-butterflies/3508594

338.  https://news.un.org/en/story/2017/10/568292-un-robot-sophia-joins-meeting-artificial-intelligence-and-sustainable

339.  https://www.yahoo.com/lifestyle/glamorous-human-like-robot-named-sophia-appears-magazine-cover-223138864.html and https://www.stylist.co.uk/people/sophia-the-robot-meaning-life-secret-happiness-exclusive-interview-five-minute-philosopher/185248

340.  https://www.bet.com/celebrities/news/2018/03/29/will-smiths-date-sophia-the-robot.html

341.  https://www.apbspeakers.com/speaker/sophia/

342.  http://www.hansonrobotics.com/robot/sophia/

343.  https://techcrunch.com/2017/10/26/saudi-arabia-robot-citizen-sophia/

344.  https://www.itbusiness.ca/news/top-canadian-researcher-says-ai-robots-deserve-human-rights/95730

345. http://www.mccormick.northwestern.edu/eecs/news/articles/2017/making-ai-systems-see-the-world-as-humans-do.html

346. http://news.mit.edu/2015/soft-robotic-hand-can-pick-and-identify-wide-array-of-objects-0930

347. https://www.digitaltrends.com/cool-tech/boston-dynamics-new-atlas-robot-best-humanoid-yet/. See also https://www.youtube.com/watch?v=8P9geWwi9e0

348. http://www.worldrecordacademy.com/technology/strongest_and_largest_robot_world_record-set_by_KUKA_70226.htm

349. http://www.ece.ucsb.edu/~ymostofi/SeeThroughImaging.html

350. https://www.theatlantic.com/technology/archive/2016/02/when-computers-started-beating-chess-champions/462216/

351. http://www.popularmechanics.com/technology/a19914/chess-computers/

352. http://www.nbcnews.com/id/19839044/ns/technology_and_science-innovation/t/checkers-computer-becomes-invincible/#.WSGgKOsrKM9

353. http://www-03.ibm.com/ibm/history/ibm100/us/en/icons/deepblue/

354. https://www.scientificamerican.com/article/how-the-computer-beat-the-go-master/

355. https://www.theatlantic.com/technology/archive/2016/03/the-invisible-opponent/475611/

356. https://www.nature.com/nature/journal/v550/n7676/full/nature24270.html

357. https://www.wired.com/2017/02/libratus/

358. https://singularityhub.com/2017/03/31/can-futurists-predict-the-year-of-the-singularity/

359. http://www.bing.com/videos/search?q=Masayoshi+Son+keynote+mobile+2017&view=detail&mid=33A84CE3BD0E5812052F33A84CE3BD0E5812052F&FORM=VIRE

360. http://www.cbsnews.com/news/60-minutes-charlie-rose-interviews-a-robot-sophia/

361.  https://qz.com/954683/sorry-humans-its-time-to-prepare-for-the-machinocene-era/

362.  https://singularityhub.com/2017/03/31/can-futurists-predict-the-year-of-the-singularity/

363.  https://bodyhackingcon.com/conference and https://www.digitaltrends.com/cool-tech/coolest-biohacking-implants/

364.  https://www.theatlantic.com/magazine/archive/2015/06/brain-hacking/392084/

365.  Many available references. See, for example, https://lifeboat.com/ex/transhumanist.technologies and http://www.thebioneer.com/cyberpunk-and-transhuman-technologies-you-can-try-right-now-or-very-soon/

366.  https://www.digitaltrends.com/cool-tech/hyundai-exoskeleton-ces-2017/

367.  http://www.pcworld.com/article/3155000/consumer-electronics/hyundais-wearable-robots-could-make-you-superhuman.html. Also see http://www.hyundainews.com/us/en/media/pressreleases/47082/hyundai-motor-leads-personal-mobility-revolution-with-advanced-wearable-robots

368.  http://www.techzone360.com/topics/techzone/articles/2017/08/18/434062-4-biohacking-facts-should-know-2017.htm and https://www.theguardian.com/technology/2017/oct/29/trans human-bodyhacking-transspecies-cyborg

369.  https://www.gofundme.com/cyborgdad

370.  http://www.lovetron.us/

371.  http://www.dailymail.co.uk/sciencetech/article-4200956/Biohacker-Rich-Lee-developing-implantable-VIBRATOR.html

372.  http://abcnews.go.com/Health/w_MindBodyNews/paralyzed-woman-moves-robotic-arm-mind/story?id=16353993

373.  http://abcnews.go.com/Health/w_MindBodyNews/paralyzed-woman-moves-robotic-arm-mind/story?id=16353993

374.  http://www.cnn.com/2006/US/03/22/btsc.oppenheim.bionic/index.html?_s=pm:us

375.   http://www.cnn.com/2006/US/03/22/btsc.oppenheim.bionic/index.html?_s=pm:us

376.   https://www.amazon.com/Can-Survive-Artificial-Intelligence-Uprising/dp/1491481250

## MASS EXTINCTION 2.0?

377.   https://www.nytimes.com/2016/12/21/upshot/the-long-term-jobs-killer-is-not-china-its-automation.html

378.   http://www.pewinternet.org/2014/08/06/future-of-jobs/

379.   https://www.wsj.com/articles/dont-fear-the-robots-1500646623

380.   https://www.mckinsey.com/~/media/McKinsey/Global%20Themes/Future%20of%20Organizations/What%20the%20future%20of%20work%20will%20mean%20for%20jobs%20skills%20and%20wages/MGI-Jobs-Lost-Jobs-Gained-Report-December-6-2017.ashx

381.   https://www.nbcnews.com/business/business-news/companies-colleges-unite-train-new-collar-students-n802251

382.   https://www-03.ibm.com/employment/us/new_collar.shtml and https://www.usatoday.com/story/tech/columnist/2016/12/13/we-need-fill-new-collar-jobs-employers-demand-ibms-rometty/95382248/

383.   https://medium.com/@marksstorm/secrets-of-silicon-valley-part-1-1bdb553fdb46

384.   http://news.rice.edu/2016/02/14/when-machines-can-do-any-job-what-will-humans-do/

385.   https://www.ft.com/content/b16e00b8-c9b7-11e0-b88b-00144feabdc0

386.   http://conexus.cberdata.org/files/MfgReality.pdf

387.   http://www.nber.org/papers/w23285 and http://www.nber.org/digest/may17/w23285.shtml

388.   http://www.sandiegouniontribune.com/opinion/sd-jobless-world-post-scarcity-world-20170308-story.html

389. http://conexus.cberdata.org/files/MfgReality.pdf

390. http://spectrum.ieee.org/automaton/robotics/industrial-robots/kinema-systems-destealths-demos-deft-depalletizer

391. https://www.youtube.com/watch?v=bDRCWnsFnC4

392. https://hypepotamus.com/news/adidas-speedfactory/

393. http://www.runningshoesguru.com/2017/06/reimagination-of-manufacturing-put-faster-sneaker-production-within-sight/

394. http://www.growingproduce.com/fruits/berries/robotic-strawberry-picker-ramping-up-for-rollout/

395. http://www.harvestcroorobotics.com/#top-of-page

396. http://www.theledger.com/news/20160727/robotic-strawberry-picker-coming-to-plant-city-company

397. https://www.reuters.com/article/us-emirates-robocop/robocop-joins-dubai-police-to-fight-real-life-crime-idUSKBN18S4K8

398. https://www.knightscope.com/

399. https://www.washingtonpost.com/news/innovations/wp/2017/12/14/crime-fighting-robot-retired-after-launching-alleged-war-on-the-homeless/?utm_term=.7847ea968528

400. https://arstechnica.com/tech-policy/2017/12/after-outcry-non-profit-stops-use-of-security-robot-to-oust-homeless/

401. https://www.wsj.com/articles/waymo-showcases-driverless-vans-without-humans-behind-the-wheel-1509433260

402. http://www.foxnews.com/auto/2017/02/16/2016-traffic-deaths-jump-to-highest-level-in-nearly-decade.html

403. https://www.nytimes.com/interactive/2018/03/20/us/self-driving-uber-pedestrian-killed.html?login=email&auth=login-email

404. https://www.bloomberg.com/news/videos/2014-10-06/sebastian-thrun-on-the-evolution-of-the-selfdriving-car

405. https://www.usatoday.com/story/money/cars/2017/11/23/self-driving-cars-programmed-decide-who-dies-crash/891493001/

406. http://www.businessinsider.com/jack-in-the-box-ceo-reconsiders-automation-kiosks-2018-1

407. http://losangeles.cbslocal.com/2017/09/13/robot-fast-food-burger/

408. https://www.npr.org/sections/thetwo-way/2018/03/05/590884388/flippy-the-fast-food-robot-sort-of-mans-the-grill-at-caliburger

409. https://techcrunch.com/2017/03/07/meet-flippy-a-burger-grilling-robot-from-miso-robotics-and-caliburger/

410. http://www.businessinsider.com/momentum-machines-funding-robot-burger-restaurant-2017-6

411. https://www.eater.com/2016/7/1/12077990/robot-burgers-san-francisco-momentum-machines

412. http://www.businessinsider.com/momentum-machines-is-hiring-2016-6

413. http://www.zdnet.com/article/lowes-introduces-autonomous-retail-service-robots/

414. https://www.theverge.com/2017/10/27/16556864/walmart-introduces-shelf-scanning-robots and http://m.arkansasonline.com/news/2017/oct/26/robots-to-work-in-50-wal-marts-20171026/

415. https://www.amazon.com/b?node=16008589011  and http://www.foxbusiness.com/features/2018/01/21/amazons-automated-grocery-store-future-opens-monday.html

416. https://www.theverge.com/2015/1/29/7939067/ap-journalism-automation-robots-financial-reporting

417. https://www.wsj.com/articles/the-robot-revolution-humanoid-potential-moving-upstream-1517221862 and http://www.adweek.com/tvnewser/this-robot-will-replace-a-tv-news-anchor-in-april/356085

418. https://www.accenture.com/us-en/insight-finance-2020-death-by-digital

419. https://www.wsj.com/articles/the-new-bookkeeper-is-a-robot-1430776272

420. https://mainichi.jp/english/articles/20161230/p2a/00m/0na/005000c

421. http://www.campaignlive.co.uk/article/why-cosabella-replaced-its-agency-ai-will-go-back-humans/1427323

422. https://www.bloomberg.com/news/articles/2017-02-28/jpmorgan-marshals-an-army-of-developers-to-automate-high-finance

423. https://www.washingtonpost.com/news/the-switch/wp/2018/02/06/algorithms-just-made-a-couple-crazy-trading-days-that-much-crazier/?utm_term=.3b3f69b75b55

424. http://money.cnn.com/2018/02/05/news/companies/dow-800-points-10-minutes/index.html

425. http://www.equbotetf.com/investor-materials/ETFMG-AIEQ-FactSheet.pdf

426. http://time.com/money/4993744/robot-mutual-fund-beating-stock-market/

427. http://www.ox.ac.uk/news/science-blog/could-oxford-developed-software-help-solve-tricky-problem-lip-reading#

428. https://theconversation.com/young-doctors-struggle-to-learn-robotic-surgery-so-they-are-practicing-in-the-shadows-89646

429. https://www.statnews.com/2018/01/10/robotic-surgery-doctors-practice/

430. http://www.davincisurgery.com/

431. http://mashable.com/2016/09/29/herosug-haptic-surgery-robot/?utm_cid=hp-n-1#q7CQhYYgAaqp

432. http://www.deakin.edu.au/about-deakin/media-releases/articles/deakin-builds-robotic-surgical-system-with-sense-of-touch

433. http://www.cyberknife.com/

434. https://www.ibm.com/blogs/watson-health/wp-content/uploads/2016/12/WHI-Overview-Executive-Brief.pdf

435. https://www.ibm.com/blogs/watson-health/wp-content/uploads/2016/12/WHI-Overview-Executive-Brief.pdf

436. https://www.ibm.com/blogs/watson-health/wp-content/uploads/2016/12/WHI-Overview-Executive-Brief.pdf

437. http://www.auntminnie.com/index.aspx?sec=ser&sub=def&pag=dis&ItemID=117841

438. https://arxiv.org/ftp/arxiv/papers/1610/1610.04662.pdf

439.   http://www.hollywoodreporter.com/behind-screen/how-artificial-intelligence-will-make-digital-humans-hollywoods-new-stars-1031553

440.   https://medium.com/@ahmed_elgammal/generating-art-by-learning-about-styles-and-deviating-from-style-norms-8037a13ae027

441.   https://medium.com/@ahmed_elgammal/generating-art-by-learning-about-styles-and-deviating-from-style-norms-8037a13ae027

442.   https://magenta.tensorflow.org/nsynth and https://www.wired.com/2017/05/google-uses-ai-create-1000s-new-musical-instruments/

443.   https://www.nytimes.com/2017/08/14/arts/design/google-how-ai-creates-new-music-and-new-artists-project-magenta.html?mcubz=0

444.   http://www.atlasobscura.com/articles/germany-robot-priest-blessu2-religion and http://www.bbc.com/news/av/world-europe-40101661/robotic-reverend-blesses-worshippers-in-eight-languages

445.   http://www.reuters.com/article/us-japan-robotpriest/in-japan-robot-for-hire-programed-to-perform-buddhist-funeral-rites-idUSKCN1B3133

446.   https://www.nytimes.com/2017/11/05/technology/machine-learning-artificial-intelligence-ai.html

447.   https://www.nytimes.com/2017/10/22/technology/artificial-intelligence-experts-salaries.html

448.   https://www.theverge.com/2017/12/5/16737224/global-ai-talent-shortfall-tencent-report and http://www.tisi.org/Public/Uploads/file/20171201/20171201151555_24517.pdf

449.   https://www.cbinsights.com/research/artificial-intelligence-startup-funding/ and http://www.slate.com/blogs/future_tense/2017/10/24/silicon_valley_is_gobbling_up_artificial_intelligence_experts.html

450. https://www.gsb.stanford.edu/insights/andrew-ng-why-ai-new-electricity

451. http://www.oxfordmartin.ox.ac.uk/downloads/academic/The_Future_of_Employment.pdf

452. http://www.mckinsey.com/global-themes/digital-disruption/harnessing-automation-for-a-future-that-works

453. http://video.foxbusiness.com/v/5719259692001/?#sp=show-clips – roughly four minutes into her interview

454. http://video.foxbusiness.com/v/5719215713001/?#sp=show-clips – roughly 3:40 into his interview

455. http://www.bbc.com/news/av/magazine-40817987/preparing-for-revolution

## MEET YOUR NEW BFF

456. https://www.bloomberg.com/news/features/2017 09-07/this-startup-is-making-virtual-people-who-look-and-act-impossibly-real

457. https://www.soulmachines.com/

458. https://www.technologyreview.com/s/603895/customer-service-chatbots-are-about-to-become-frighteningly-realistic/

459. https://www.bloomberg.com/news/features/2017-09-07/this-startup-is-making-virtual-people-who-look-and-act-impossibly-real

460. http://www.irobot.com/

461. http://newatlas.com/honda-miimo-robot-lawn-mower-us/48994/

462. https://www.amazon.com/p/feature/ofoyqn7wjy2p39a and https://www.amazon.com/Amazon-Echo-Bluetooth-Speaker-with-WiFi-Alexa/dp/B00X4WHP5E?&_encoding=UTF8&tag=biipg_070617_best-smart-speakers-20&linkCode=ur2&linkId=716d141e95e7a4edc8cc197c4ad10952&camp=1789&creative=9325#tech

463. http://www.businessinsider.com/best-smart-speaker-amazon-echo/#the-best-affordable-smart-speaker-3

464. https://www.linkedin.com/pulse/amazon-echo-magical-its-also-turning-my-kid-asshole-hunter-walk?trk=Inc

465. https://www.amazon.com/Echo-Hands-Free-Camera-Style-Assistant/dp/B0186JAEWK?tag=bisafetynet2-20

466. https://www.amazon.com/Amazon-Echo-Show-Alexa-Enabled-Black/dp/B01J24C0TI/ref=sr_1_1?s=amazon-devices&ie=UTF8&qid=1515945570&sr=1-1&keywords=echo+show

467. https://madeby.google.com/home/

468. https://www.apple.com/homepod/

469. http://www.harmankardon.com/invoke.html

470. https://martechtoday.com/facebook-shutters-digital-assistant-m-readies-video-portal-209233

471. http://time.com/money/4874972/kuri-robot-800-dollars/ and https://techcrunch.com/2017/06/21/home-robot-kuri-can-now-recognize-pets-see-and-stream-in-hd/

472. https://zenbo.asus.com/product/overview/

473. https://www.theverge.com/2016/5/30/11810102/asus-zenbo-home-helper-robot

474. https://www.ald.softbankrobotics.com/en/robots/pepper

475. https://www.wired.com/2016/06/pepper-emotional-robot-learns-feel-like-american/

476. https://www.theverge.com/2016/8/12/12453770/pepper-emotional-robot-softbank-b8ta-silicon-valley

477. https://www.wired.com/2016/06/pepper-emotional-robot-learns-feel-like-american/

478. https://theloupe.io/pepper-the-humanoid-robot-f88c09774dc6 and https://wsimag.com/science-and-technology/19093-a-sprinkle-of-pepper

479. http://www.worldatlas.com/articles/countries-with-the-largest-aging-population-in-the-world.html

480. https://qz.com/217199/softbanks-humanoid-robot-will-be-great-for-tending-to-japans-elderly/

481. https://realbotix.systems/#products
482. http://www.foxnews.com/tech/2017/02/06/realdoll-builds-artificially-intelligent-sex-robots-with-programmable-personalities.html
483. https://realbotix.systems/#products
484. https://www.howardstern.com/news/2017/8/10/video-meet-harmony-sexbot-realest-realdoll-ever-assembled/
485. https://www.scientificamerican.com/article/humans-marrying-robots/
486. http://www.mirror.co.uk/news/weird-news/sex-robots-fully-functional-genitalia-9161419?service=responsive
487. https://campaignagainstsexrobots.org/
488. https://www.nytimes.com/2017/07/17/opinion/sex-robots-consent.html?mcubz=0
489. https://responsiblerobotics.org/2017/07/05/frr-report-our-sexual-future-with-robots/
490. https://www.thecollegefix.com/post/35300/
491. https://lumidolls.com/collections/rent-sex-doll
492. https://www.siliconwives.com/blogs/news/shut-down-lumidolls-sex-doll-brothel-in-barcelona-closes-doors
493. http://www.cetusnews.com/news/World-s-first-sex-doll-brothel-to-GO-GLOBAL-with-new-venues-in-US-and-across-Europe.S1NUMobA33yM.html
494. http://www.reuters.com/article/us-afghanistan-drones-exclusive/exclusive-afghan-drone-war-data-show-unmanned-flights-dominate-air-campaign-idUSKCN0XH2UZ
495. http://dronecenter.bard.edu/underwater-drones/
496. https://www.washingtonpost.com/news/business/wp/2017/09/15/robot-submarines-could-soon-be-used-to-spy-on-americas-enemies/?utm_term=.180eb3b94da9
497. https://www.defense.gov/Portals/1/Documents/pubs/Perdix%20Fact%20Sheet.pdf
498. https://www.cbsnews.com/news/60-minutes-autonomous-drones-set-to-revolutionize-military-technology/

499. https://www.defense.gov/News/News-Releases/News-Release-View/Article/1044811/department-of-defense-announces-successful-micro-drone-demonstration/

500. http://breakingdefense.com/2016/09/killer-robots-never-says-defense-secretary-carter/

501. http://newatlas.com/kalashnikov-ai-weapon-terminator-conundrum/50576/

502. https://www.defensenews.com/2016/01/23/the-terminator-conundrum-pentagon-weighs-ethics-of-pairing-deadly-force-ai/

503. https://www.nytimes.com/2016/10/26/us/pentagon-artificial-intelligence-terminator.html?mcubz=0

504. https://futureoflife.org/autonomous-weapons-open-letter-2017/

## LOOK! IT'S A HUMAN. IT'S A MACHINE. IT'S A . . .

505. http://www.dailymail.co.uk/sciencetech/article-4319436/Singularity-create-super-humans-Google-expert-claims.html

506. http://www.bbc.com/news/technology-30290540

507. http://www.techrepublic.com/article/transhumanism-should-we-use-robotics-to-enhance-humans/

508. http://www.imdb.com/title/tt0071054/plotsummary?ref_=tt_stry_pl

509. https://www.nbcnews.com/mach/technology/godlike-homo-deus-could-replace-humans-tech-evolves-n757971

510. https://www.nbcnews.com/mach/technology/godlike-homo-deus-could-replace-humans-tech-evolves-n757971

511. https://www.nbcnews.com/mach/technology/godlike-homo-deus-could-replace-humans-tech-evolves-n757971

512. https://www.wsj.com/articles/a-hardware-update-for-the-human-brain-1496660400

513. https://www.reuters.com/article/us-germany-bookfair-dan-brown/collective-consciousness-to-replace-god-author-dan-brown-idUSKBN1CH1O1

514. http://dilbert.com/strip/1996-02-11

515. http://www.nature.com/nnano/journal/v10/n7/full/nnano.2015
.115.html#author-information

516. https://www.frontiersin.org/articles/10.3389/fnhum.2016.00034/
full. Also, for example, see https://www.sott.net/article/315145-
Breakthrough-technology-or-hype-Scientists-develop-program-
to-upload-skills-and-knowledge-directly-to-the-brain and https://
thehackernews.com/2016/03/upload-skills-to-brain.html

517. https://twitter.com/clonmusk/status/739006012749799424
?lang=en

518. http://www.secondsight.com/how-is-argus-r-ii-designed-to-
produce-sight-en.html

519. http://www.bbc.com/future/story/20140923-im-blind-but-i-have-
bionic-eyes

520. https://www.osa-opn.org/home/articles/volume_28/april_2017/
features/vision_accomplished_the_bionic_eye/

521. https://www.technologyreview.com/s/608844/blind-patients-to-
test-bionic-eye-brain-implants/

522. https://32market.com/public/

523. https://www.cnbc.com/2017/08/11/three-square-market-ceo-
explains-its-employee-microchip-implant.html

524. https://www.usatoday.com/story/tech/2017/08/09/you-get-
chipped-eventually/547336001/

525. https://www.cnbc.com/2017/08/11/three-square-market-ceo-
explains-its-employee-microchip-implant.html

526. http://foreignpolicy.com/2009/10/23/transhumanism/

527. http://foreignpolicy.com/2009/10/23/transhumanism/

528. http://www.dailymail.co.uk/sciencetech/article-4319436/
Singularity-create-super-humans-Google-expert-claims.html

529. http://foreignpolicy.com/2009/10/23/transhumanism/

530. https://www.nbcnews.com/mach/technology/godlike-homo-deus-
could-replace-humans-tech-evolves-n757971

531. https://www.technologyreview.com/s/527336/do-we-need-
asimovs-laws/

532. https://www.washingtonpost.com/opinions/matt-miller-artificial-intelligence-our-final-invention/2013/12/18/26ed6be8-67e6-11e3-8b5b-a77187b716a3_story.html?utm_term=.0745a02e24c7
533. http://www.bbc.co.uk/programmes/b0916ghz
534. http://www.bbc.co.uk/programmes/b0916ghz
535. See, for example: http://www.spiked-online.com/spiked-review/article/the-new-masters-of-the-universe/17684 and http://www.spiegel.de/international/germany/spiegel-cover-story-how-silicon-valley-shapes-our-future-a-1021557.html
536. Harari, Yuval Noah. 2017. Chapter 10. In Homo Deus: A Brief History of Tomorrow. 356. New York: Harper.
537. https://www.buzzfeed.com/alexkantrowitz/were-in-an-artificial-intelligence-hype-cycle?utm_term=.rqy2MX2Z5#.jpzeJve7V
538. http://aeroastro.mit.edu/videos/centennial-symposium-one-one-one-elon-musk
539. http://www.sfchronicle.com/aboutsfgate/article/Why-universal-basic-income-is-gaining-support-11290211.php
540. http://www.businessinsider.com/y-combinator-basic-income-test-2017-9
541. https://www.cnbc.com/2017/09/21/silicon-valley-giant-y-combinator-to-branch-out-basic-income-trial.html
542. https://ww2.kqed.org/news/2018/01/22/stockton-gets-ready-to-experiment-with-universal-basic-income/
543. https://www.usatoday.com/story/opinion/2017/08/11/cutting-welfare-helps-people-christian-schneider-column/557082001/
544. http://workandhappinessfilm.com/

# SPY

## MEMORY LANE

545. http://www.jareddiamond.org/Jared_Diamond/The_World_Until_Yesterday.html

546. https://www.amazon.com/dp/B01CGHXKZG?ref_=kcr_store_sample&tag=x_gr_w_preview_kcr-20&at=x_gr_w_preview_kcr-20

547. https://tspace.library.utoronto.ca/bitstream/1807/24728/1/Cooley_Alice_J_201002_PhD_thesis.pdf p. 19

548. https://books.google.com/books/about/Night_in_the_Middle_Ages.html?id=zV-BAAAAMAAJ

549. https://www.amazon.com/History-Private-Life-Revelations-Paperback/dp/0674400011/ref=pd_bxgy_14_2?_encoding =UTF8&pd_rd_i=0674400011&pd_rd_r=3XAF26ZA5 F8BHBX2PDNS&pd_rd_w=O2eHO&pd_rd_wg=xIKrL&psc= 1&refRID=3XAF26ZA5F8BHBX2PDNS

550. https://www.archives.gov/founding-docs/declaration-transcript

551. https://www.archives.gov/founding-docs/constitution-transcript

552. https://www.amazon.com/dp/B01CGHXKZG?ref_=kcr_store_sample&tag=x_gr_w_preview_kcr-20&at=x_gr_w_preview_kcr-20

553. 1989. The Practical Impact of Writing. In A History of Private Life III: Passions of the Renaissance, 111. Cambridge, MP: Belknap Press, Cambridge, MA.

554. http://faculty.uml.edu/sgallagher/Brandeisprivacy.htm

555. https://techcrunch.com/2013/11/20/googles-cerf-says-privacy-may-be-an-anomaly-historically-hes-right/

556. https://www.newyorker.com/magazine/2013/06/24/the-prism

557. http://www.nola.com/business/index.ssf/2010/01/early_telephone_operator_recal.html

558. http://www.lyncmigration.com/news/2013/06/13/7202216.htm

559. https://www.newyorker.com/magazine/2013/06/24/the-prism

560. https://www.risapuno.com/please-enable-cookies/

561. https://www.propublica.org/article/how-much-of-your-data-would-you-trade-for-a-free-cookie

562. https://www.theatlantic.com/technology/archive/2015/02/why-people-probably-wont-pay-to-keep-their-web-history-secret/385765/

563. https://www.theatlantic.com/technology/archive/2015/02/why-people-probably-wont-pay-to-keep-their-web-history-secret/385765/

564. http://www.businessinsider.com/snowden-leaks-timeline-2016-9

565. http://www.abc.net.au/news/2017-03-09/are-all-these-data-breaches-down-to-the-edward-snowden-effect/8335962

566. http://www.bbc.com/news/technology-39197664

567. https://medium.com/the-ferenstein-wire/the-birth-and-death-of-privacy-3-000-years-of-history-in-50-images-614c26059e

568. https://www.goodreads.com/quotes/916551-il-n-est-si-homme-de-bien-qu-il-mette-l-examen

## LIVING IN A FISHBOWL

569. https://www.youtube.com/watch?v=rXwdnHnLvms

570. https://www.wsj.com/articles/chinas-new-tool-for-social-control-a-credit-rating-for-everything-1480351590

571. https://www.nec.com/en/global/ad/smartcity/face/

572. http://best-practices.frost-multimedia-wire.com/nec

573. http://www.eenewseurope.com/news/face-recognition-module-operates-real-time-crowd-video-capture

574. https://www.necam.com/Docs/?id=09841596-da09-4f39-8c03-5ce9e81e8c30

575. http://affect.media.mit.edu/

576. https://www.affectiva.com/

577. https://www.reuters.com/article/us-emotient-m-a-apple/apple-buys-artificial-intelligence-startup-emotient-wsj-idUSKBN0UL27420160107

578. https://www.wsj.com/articles/who-wants-to-supply-chinas-surveillance-state-the-west-1509540111

579. https://www.wsj.com/articles/apple-buys-artificial-intelligence-startup-emotient-1452188715

580. https://www.beijing-kids.com/blog/2017/04/29/could-baidus-cross-age-facial-recognition-ai-help-end-child-abductions/

581. https://www.wsj.com/articles/who-wants-to-supply-chinas-surveillance-state-the-west-1509540111

582. https://www.forbes.com/sites/ywang/2017/07/11/how-china-is-quickly-embracing-facial-recognition-tech-for-better-and-worse/#320756126856

583. https://www.wsj.com/articles/the-all-seeing-surveillance-state-feared-in-the-west-is-a-reality-in-china-1498493020

584. https://www.prnewswire.com/news-releases/china-ping-an-launches-worlds-first-face-recognition-loan-technology-300252820.html

585. https://www.wsj.com/articles/the-all-seeing-surveillance-state-feared-in-the-west-is-a-reality-in-china-1498493020 and https://www.theguardian.com/technology/2017/jan/11/china-beijing-first-smart-restaurant-kfc-facial-recognition

586. https://www.faceplusplus.com/face-detection/

587. https://www.wsj.com/articles/the all seeing-surveillance-state-feared-in-the-west-is-a-reality-in-china-1498493020 and https://www.businessinsider.com.au/china-facial-recognition-technology-works-in-one-second-2018-3

588. http://www.chinadaily.com.cn/china/2015-10/05/content_22091634.htm and https://www.theguardian.com/world/2017/mar/20/face-scanners-public-toilet-tackle-loo-roll-theft-china-beijing

589. http://en.people.cn/n3/2018/0326/c90000-9441798.html

590. https://www.wsj.com/articles/the-all-seeing-surveillance-state-feared-in-the-west-is-a-reality-in-china-1498493020

591. http://www.chinadaily.com.cn/china/2017-09/25/content_32466168.htm

592. http://www.scmp.com/tech/china-tech/article/2138960/jaywalkers-under-surveillance-shenzhen-soon-be-punished-text

593. https://www.wsj.com/articles/chinese-police-go-robocop-with-facial-recognition-glasses-1518004353?mod=searchresults&page=2&pos=7

594. https://www.engadget.com/2017/05/10/baidu-ai-facial-recognition-finds-abducted-son/

595. http://www.chinaaid.org/2017/04/zhejiang-province-orders-churches-to.html

596. http://www.scmp.com/news/china/policies-politics/article/2084169/chinas-jerusalem-anti-terror-cameras-new-cross-churches

597. https://www.yahoo.com/news/iphone-brings-face-recognition-fears-masses-012915239.html

598. https://www.reuters.com/article/us-apple-iphone-privacy-analysis/app-developer-access-to-iphone-x-face-data-spooks-some-privacy-experts-idUSKBN1D20DZ

599. http://www.miamiherald.com/news/nation-world/national/article153911364.html

600. http://www.nextgov.com/emerging-tech/2017/11/dhs-wants-tech-scan-your-face-you-drive-mexico/142218/

601. https://www.fbi.gov/services/cjis/fingerprints-and-other-biometrics/ngi

602. https://www.nbcnews.com/news/us-news/facial-recognition-databases-hearing-congress-hammers-fbi-n737461

603. https://www.fastcompany.com/3069264/congress-fbi-face-recognition-real-time-street-lineup

604. https://www.nbcnews.com/news/us-news/facial-recognition-databases-hearing-congress-hammers-fbi-n737461

605. https://www.perpetuallineup.org/

606. https://www.fastcompany.com/3069264/congress-fbi-face-recognition-real-time-street-lineup

607. https://www.fastcompany.com/3069264/congress-fbi-face-recognition-real-time-street-lineup

608. https://www.wsj.com/articles/the-all-seeing-surveillance-state-feared-in-the-west-is-a-reality-in-china-1498493020

609. http://www.worldometers.info/world-population/us-population/ and http://www.worldometers.info/world-population/china-population/

610. https://ipvm.com/reports/america-cctv-recording
611. http://sec.kingston.ac.uk/news/2012/cutting-edge-advance-in-cctv-software-earns-industry-acclaim/
612. https://ring.com/security-cams
613. https://nest.com/cameras/
614. http://www.nkytribune.com/2017/04/keven-moore-surveillance-cameras-are-everywhere-providing-protection-but-not-much-privacy/
615. http://www.worldatlas.com/articles/most-spied-on-cities-in-the-world.html
616. https://www.cctv.co.uk/how-many-cctv-cameras-are-there-in-london/ and https://www.caughtoncamera.net/news/how-many-cctv-cameras-in-london/
617. http://ukpopulation2018.com/population-of-london-2018.html
618. http://populationof2017.com/population-of-beijing-2017.html
619. https://www.thenewamerican.com/usnews/crime/item/22285 brother-watching-you-when-spy-cameras-are-everywhere
620. https://www.cityoflondon.police.uk/advice-and-support/safer-roads/Pages/ANPR.aspx
621. http://www.independent.co.uk/news/uk/this-britain/how-average-briton-is-caught-on-camera-300-times-a-day-5354728.html
622. http://www.businessinsider.com/despite-being-the-most-watched-city-in-the-world-london-is-no-safer-for-all-its-cctv-2012-2 and https://bigbrotherwatch.org.uk/
623. http://www.businessinsider.com/despite-being-the-most-watched-city-in-the-world-london-is-no-safer-for-all-its-cctv-2012-2
624. https://epic.org/2013/10/epic-foia---fbi-says-20-error-.html
625. http://journals.plos.org/plosone/article?id=10.1371/journal.pone.0139827
626. https://www.fastcompany.com/3069264/congress-fbi-face-recognition-real-time-street-lineup
627. http://www.gao.gov/products/GAO-16-267
628. https://www.washingtonpost.com/news/the-switch/wp/2016/06/16/you-could-be-in-this-fbi-facial-recognition-system-

and-not-even-know-it/?utm_term=.25dea3cee7c8 and https://arstechnica.com/tech-policy/2016/06/smile-youre-in-the-fbi-face-recognition-database/

629. https://www.justice.gov/opcl/privacy-act-1974
630. https://www.pindrop.com/blog/fbi-biometric-database-will-exempt-privacy-act-protections/
631. https://www.federalregister.gov/documents/2017/08/01/2017-15423/privacy-act-of-1974-implementation
632. https://www.federalregister.gov/documents/2017/08/01/2017-15423/privacy-act-of-1974-implementation
633. https://cyber.harvard.edu/
634. Schneier, Bruce. 2015. *Secrets and Lies: Digital Security in a Networked World* Hoboken, NJ: Wiley.

## OMNISCIENT OBJECTS OF OUR DESIRE

635. http://iotlist.co/
636. https://www.gartner.com/newsroom/id/3598917
637. http://www.zdnet.com/article/iot-devices-will-outnumber-the-worlds-population-this-year-for-the-first-time/
638. https://www.gartner.com/newsroom/id/3598917
639. https://www.shodan.io/
640. https://www.wired.com/2012/02/home-cameras-exposed/ and https://www.hackread.com/website-streams-from-private-security-cameras/
641. https://www.washingtonpost.com/news/the-switch/wp/2017/03/07/why-the-cia-is-using-your-tvs-smartphones-and-cars-for-spying/?utm_term=.196464d7dae3
642. https://www.washingtonpost.com/news/the-switch/wp/2017/03/07/why-the-cia-is-using-your-tvs-smartphones-and-cars-for-spying/?utm_term=.196464d7dae3
643. http://www.mcclatchydc.com/news/nation-world/national/national-security/article166488597.html

644. http://www.mcclatchydc.com/news/nation-world/national/
national-security/article166488597.html

645. http://icitech.org/

646. http://www.mcclatchydc.com/news/nation-world/national/
national-security/article166488597.html

647. http://www.mcclatchydc.com/news/nation-world/national/
national-security/article166488597.html

648. https://www.statista.com/chart/12455/number-of-apps-available-
in-leading-app-stores/

649. https://www.wired.com/2017/05/hundreds-apps-can-listen-
beacons-cant-hear/

650. https://www.silverpush.co/home and http://christian.wressnegger.
info/content/projects/sidechannels/2017-eurosp.pdf and https://
thehackernews.com/2017/05/ultrasonic-tracking-signals-apps
.html

651. https://www.silverpush.co/about#

652. https://www.wired.com/2016/11/block-ultrasonic-signals-didnt-
know-tracking/

653. http://christian.wressnegger.info/content/projects/sidechannels/
2017-eurosp.pdf

654. https://www.bleepingcomputer.com/news/security/234-android-
applications-are-currently-using-ultrasonic-beacons-to-track-
users/

655. https://www.digitaltrends.com/mobile/is-your-smartphone-
listening-to-your-conversations/

656. https://www.reuters.com/article/us-irobot-strategy/roomba-
vacuum-maker-irobot-betting-big-on-the-smart-home-
idUSKBN1A91A5

657. https://www.google.com/search?q=amazon+roomba&client=fire
fox-b-1&source=univ&tbm=shop&tbo=u&sa=X&ved=0ahUK
EwiJi-H7rN_YAhUESK0KHXJ5DWEQsxgIKA&biw=1600&
bih=767

658.    http://www.reuters.com/article/us-irobot-strategy/roomba-vacuum-maker-irobot-betting-big-on-the-smart-home-idUSKBN1A91A5

659.    http://www.zdnet.com/article/exclusive-roomba-ceo-responds-to-spying-story/

660.    http://www.reuters.com/article/us-irobot-strategy/roomba-vacuum-maker-irobot-betting-big-on-the-smart-home-idUSKBN1A91A5

661.    https://www.bose.com/en_us/products/headphones/wireless_headphones.html

662.    https://www.scribd.com/document/345620278/Bose-Privacy-Complaint?irgwc=1&content=10079&campaign=Skimbit%2C%20Ltd.&ad_group=58287X1517249X7d8ee3538000f5d50e8b218b414a78b6&keyword=ft750noi&source=impactradius&medium=affiliate

663.    https://downloads.bose.com/ced/bose_connect/privacy_policy.html

664.    https://www.scribd.com/document/345620278/Bose-Privacy-Complaint?irgwc=1&content=10079&campaign=Skimbit%2C%20Ltd.&ad_group=58287X1517249X7d8ee3538000f5d50e8b218b414a78b6&keyword=ft750noi&source=impactradius&medium=affiliate

665.    https://segment.com/

666.    https://www.nuance.com/mobile/automotive/dragon-drive.html

667.    https://www.ford.com/technology/sync/

668.    http://www.gmc.com/intellilink-infotainment-system.html

669.    http://smarthome.reviewed.com/best-right-now/the-best-baby-audio-monitors

670.    https://www.fitbit.com/home

671.    https://www.apple.com/apple-watch-series-3/?afid=p238%7Cs2QbmM6cp-dc_mtid_1870765e38482_pcrid_224100368864_&cid=aos-us-kwgo-watch--slid--product-

672.    https://www.amazon.com/s/?ie=UTF8&keywords=cowatch&tag=googhydr-20&index=aps&hvadid=153671799815&hvpo

s=1t3&hvnetw=g&hvrand=8367606284507894858&hvpone=
&hvptwo=&hvqmt=e&hvdev=c&hvdvcmdl=&hvlocint=&hvlo
cphy=9013202&hvtargid=kwd-12116419788&ref=pd_sl_68od0
vcodk_e

673. https://www.amazon.com/dp/B07456BG8N?tag=googhydr-
20&hvadid=223605111254&hvpos=1t1&hvnetw=g&hvrand=3
934109330685590374&hvpone=&hvptwo=&hvqmt=e&hvdev
=c&hvdvcmdl=&hvlocint=&hvlocphy=9013202&hvtargid=kwd-
295921611050&ref=pd_sl_2g7cb1h5ze_e

674. https://store.google.com/product/google_home

675. https://www.apple.com/homepod/?afid=p238%7Cs3C6QO
c9h-dc_mtid_20925u8r61652_pcrid_228934983503&cid=wwa-
us-kwgo-aes-slid--

676. https://www.theverge.com/2017/1/7/14200210/amazon-alexa-
tech-news-anchor-order-dollhouse

677. http://www.foxnews.com/tech/2017/01/03/6-year-old-
accidentally-orders-high-end-treats-with-amazons-alexa.html

678. http://www.cnn.com/2017/01/06/tech/alexa-dollhouses-san-
diego-irpt-trnd/index.html

679. https://www.wired.com/2016/12/alexa-and-google-record-your-
voice/

680. https://www.ic3.gov/media/2017/170717.aspx

681. http://cloudpets.com/how-it-works

682. https://motherboard.vice.com/en_us/article/pgwean/internet-of-
things-teddy-bear-leaked-2-million-parent-and-kids-message-
recordings and http://money.cnn.com/2017/02/27/technology/
cloudpets-data-leak-voices-photos/index.html and http://www
.huffingtonpost.com.au/2017/02/28/millions-of-private-messages-
between-parents-and-kids-hacked-in_a_21816860/

683. https://www.troyhunt.com/data-from-connected-cloudpets-teddy-
bears-leaked-and-ransomed-exposing-kids-voice-messages/

684. http://money.cnn.com/2017/02/27/technology/cloudpets-data-
leak-voices-photos/index.html

685. http://www.bbc.com/news/technology-39115001

686.   https://www.networkworld.com/article/3175225/security/smart-teddy-bears-involved-in-a-contentious-data-breach.html

687.   http://www.newsweek.com/meet-hello-barbie-wi-fi-doll-talks-children-307482 and http://www.slate.com/blogs/future_tense/2017/07/19/fbi_is_warning_parents_about_hacking_internet_connected_toys.html and https://www.geektime.com/2017/07/20/fbi-warns-parents-internet-connected-toys-can-spy-on-your-kids/

688.   https://www.ic3.gov/media/2017/170717.aspx

689.   http://we-vibe.com/app

690.   http://we-vibe.com/app

691.   https://motherboard.vice.com/en_us/article/bmv5ja/a-sex-toy-lawsuit-highlights-privacy-concerns-around-smart-dildos

692.   https://www.documentcloud.org/documents/3106576-NP-v-Standard-Innovation-Complaint.html

693.   https://www.pentestpartners.com/ and https://motherboard.vice.com/en_us/article/bmv5ja/a-sex-toy-lawsuit-highlights-privacy-concerns-around-smart-dildos

694.   https://www.scribd.com/document/341529556/We-Vibe-Settlement

695.   http://www.debate.org/debates/Should-we-be-tracked-on-internet/1/

696.   http://www.breitbart.com/tech/2017/04/14/report-school-owned-devices-spying-on-kids/

697.   https://www.engadget.com/2016/10/21/mirai-botnet-hacked-cameras-routers-internet-outage/

698.   https://blog.checkpoint.com/2017/10/19/new-iot-botnet-storm-coming/

699.   http://www-03.ibm.com/ibm/history/ibm100/us/en/icons/smarterplanet/

700.   https://cryptome.org/2012/12/assange-crypto-arms.htm

## OPERATING IN THE SHADOWS

701. https://tspace.library.utoronto.ca/bitstream/1807/24728/1/Cooley_Alice_J_201002_PhD_thesis.pdf p.16

702. http://www.drroberting.com/articles/holmes.pdf

703. http://www.drroberting.com/articles/holmes.pdf

704. http://www.forensictv.net/Downloads/forensic_science/forensic_science_timeline_by_norah_rudin_and_keith_inman.pdf

705. http://scholarship.law.edu/cgi/viewcontent.cgi?article=1018&context=jlt

706. https://www.youtube.com/watch?v=LiZH5eH5eDw#action=share

707. https://www.youtube.com/watch?v=D0pBbF1AR8s

708. https://www.youtube.com/watch?v=LiZH5eH5eDw#action=share and https://www.youtube.com/watch?v=QR5nn6HGl1w

709. https://www.youtube.com/watch?v=C5wUAaIdF6c

710. http://insider.foxnews.com/2016/12/12/orange-coast-college-teachers-union-legal-action-student-professor-anti-trump-rant

711. https://www.washingtonpost.com/news/grade-point/wp/2017/02/15/a-professor-called-trumps-victory-terrorism-a-student-who-recorded-the-rant-got-suspended/?utm_term=.4edd47d30db8

712. http://losangeles.cbslocal.com/2016/12/08/orange-coast-college-student-threatened-with-expulsion-after-recording-professors-anti-trump-tirade/

713. http://www.ocregister.com/2017/02/24/occ-rescinds-suspension-of-student-who-recorded-teachers-anti-trump-comments/

714. https://www.travelchinaguide.com/attraction/beijing/heaven/

715. http://www.euronews.com/2017/03/20/beijing-park-rolls-out-toilet-tissue-security

716. https://weibo.com/ttarticle/p/show?id=2309404087058741217036

717. http://www.euronews.com/2017/03/20/beijing-park-rolls-out-toilet-tissue-security

718. https://www.theguardian.com/world/2017/mar/20/face-scanners-public-toilet-tackle-loo-roll-theft-china-beijing

719. https://www.fbi.gov/services/laboratory/biometric-analysis/codis/ndis-statistics

720. https://snapshot.parabon-nanolabs.com/#phenotyping

721. https://www.genome.gov/pages/education/dnaday/animations/makingsnpsmakesense.html

722. https://snapshot.parabon-nanolabs.com/#phenotyping

723. https://snapshot.parabon-nanolabs.com/#phenotyping-how

724. http://www.nationalgeographic.com/magazine/2016/07/forensic-science-justice-crime-evidence/

725. See also https://www.forensicmag.com/news/2017/07/tip-dna-phenotype-snapshot-leads-arrest-2009-murder-cold-case and https://www.nationalgeographic.com/magazine/2016/07/forensic-science-justice-crime-evidence/

726. https://www.cpso.com/2017/07/24/cpso-makes-arrest-in-bouzigard-murder/ and http://www.americanpress.com/news/crime/dequincy-man-accused-in-slaying/article_0cfd33fc-7143-11e7-9634-4fba46e8d95d.html

727. http://www.cnn.com/2017/07/10/us/alexa-calls-police-trnd/index.html

728. http://mashable.com/2017/07/10/amazon-echo-calls-sheriff/#MbBmO7AO.sqg

729. http://abcnews.go.com/US/smart-home-device-alerts-mexico-authorities-alleged-assault/story?id=48470912

730. https://www.washingtonpost.com/local/public-safety/commit-a-crime-your-fitbit-key-fob-or-pacemaker-could-snitch-on-you/2017/10/09/f35a4f30-8f50-11e7-8df5-c2e5cf46c1e2_story.html?utm_term=.3d3d9125caf0

731. http://people.com/crime/friend-speaks-connie-debate-fitbit-murder-by-husband-richard/

732. https://www.washingtonpost.com/local/public-safety/commit-a-crime-your-fitbit-key-fob-or-pacemaker-could-snitch-on-

you/2017/10/09/f35a4f30-8f50-11e7-8df5-c2e5cf46c1e2_story
.html?utm_term=.1474a9f8ff28

733.   https://www.fitbit.com/home

734.   https://patch.com/connecticut/ellington-somers/ellington-murder-
case-gets-another-continuance

735.   http://www.courant.com/news/connecticut/hc-news-dabate-fit-
bit-murder-continued-20180119-story.html

736.   https://www.goodreads.com/quotes/206413-all-human-beings-
have-three-lives-public-private and-secret

737.   https://www.bloomberg.com/profiles/companies/VSNX:US-
visionics-corp

738.   https://www.appliancedesign.com/articles/83587-visionics-
announces-faceit-argus-10-3

739.   http://www.nytimes.com/2001/10/07/magazine/a-watchful-state.
html

740.   https://www.aclu.org/blog/privacy-technology/surveillance-
technologies/eight-problems-police-threat-scores

741.   http://www.fresnobee.com/opinion/readers-opinion/article
60644891.html

742.   https://www.aclu.org/blog/privacy-technology/surveillance-
technologies/eight-problems-police-threat-scores

743.   http://www.intrado.com/ and https://beware.intrado.com/#!login

744.   https://supreme.justia.com/cases/federal/us/385/323/case.html

745.   https://www.theverge.com/2017/10/30/16570148/neuroscience-
suicide-mental-health-brain-imagining-fmri-machine-learning-
algorithm

746.   http://neurosciencenews.com/machine-learning-thought-6974/

747.   http://neurosciencenews.com/machine-learning-thought-6974/

748.   http://www.cnn.com/2014/02/04/tech/innovation/this-new-tech-
can-detect-your-mood/index.html

749.   https://www.goodreads.com/work/quotes/14826-the-maytrees

# FRANKENSTEIN

## MEMORY LANE

750. https://books.google.com/books/about/The_Origin_of_Species_by_Means_of_Natura.html?id=6AUAAAAAQAAJ&printsec=frontcover&source=kp_read_button#v=onepage&q&f=false

751. For example, see https://courses.lumenlearning.com/boundless-biology/chapter/population-genetics/

752. https://www.acf.org/the-american-chestnut/native-range-map/

753. http://www.pressherald.com/2017/08/27/saving-an-american-classic/  and https://www.npr.org/sections/thesalt/2012/12/13/167142393/sowing-the-seeds-for-a-great-american-chestnut-comeback

754. http://www.pressherald.com/2017/08/27/saving-an-american-classic/

755. https://bangordailynews.com/2015/11/28/news/state/tallest-chestnut-tree-in-north-america-found-in-maine/

756. http://mainepublic.org/post/tallest-american-chestnut-tree-found-maine#stream/0

757. http://mainepublic.org/post/tallest-american-chestnut-tree-found-maine#stream/0

758. http://www.azquotes.com/quote/896780

759. https://www.smithsonianmag.com/travel/why-wild-tiny-pimp-tomato-so-important-180955911/

760. https://aggie-horticulture.tamu.edu/archives/parsons/publications/vegetabletravelers/tomato.html

761. https://aggie-horticulture.tamu.edu/archives/parsons/publications/vegetabletravelers/tomato.html

762. https://www.the-scientist.com/?articles.view/articleNo/41194/title/360-Degree-View-of-the-Tomato/

763. http://www.pnas.org/content/106/13/4957.full#ref-1

764. http://www.nytimes.com/2010/05/25/science/25creature.html

765. http://learn.genetics.utah.edu/content/selection/corn/

766. http://www.nytimes.com/2010/05/25/science/25creature.html

767. https://passel.unl.edu/pages/informationmodule.php?idinformat
ionmodule=1075412493&topicorder=2&maxto=12&minto=1

768. https://www.nature.com/articles/ncomms16082

769. http://journals.plos.org/plosone/article?id=10.1371/journal
.pone.0022821

770. http://www.fci.be/en/Nomenclature/Default.aspx

771. http://wallace.genetics.uga.edu/groups/evol3000/wiki/ce8b9/
Selective_Breeding_or_Artificial_Selection.html

772. http://www.chronicle.com/blogs/brainstorm/the-mendel-darwin-
connection/24433

773. https://www.biography.com/people/gregor-mendel-39282

774. http://www.pbs.org/wgbh/evolution/library/06/1/l_061_01.html

775. https://www.smithsonianmag.com/arts-culture/evolution-world-
tour-mendels-garden-czech-republic-6088291/

776. https://www.smithsonianmag.com/arts-culture/evolution-world-
tour-mendels-garden-czech-republic-6088291/

777. http://www.bbc.co.uk/schools/gcsebitesize/science/edexcel_
pre_2011/genes/genesrev1.shtml

778. https://health.utah.gov/genomics/familyhistory/documents/
Family%20Reunion/reference%20guide.pdf

779. https://www.nature.com/scitable/content/this-1-1-1-1-phenotypic-
ratio-6888599

780. https://www.nature.com/scitable/topicpage/discovery-of-dna-
structure-and-function-watson-397

781. https://www.dna-worldwide.com/resource/160/history-dna-
timeline

782. https://profiles.nlm.nih.gov/SC/Views/Exhibit/narrative/
doublehelix.html

783. https://www.genome.gov/11006943/human-genome-project-
completion-frequently-asked-questions/

784. http://abcnews.go.com/GMA/story?id=2192678&page=1

785. https://gizmodo.com/heres-why-todays-decision-over-who-
invented-crispr-matt-1792402072

## NOT OF THIS WORLD

786. http://articles.latimes.com/1992-06-04/food/fo-1061_1_plant-breeding
787. http://articles.latimes.com/1992-06-04/food/fo-1061_1_plant-breeding
788. http://www.nytimes.com/2013/06/24/booming/you-call-that-a-tomato.html
789. http://www.nytimes.com/1994/05/19/us/tomato-review-no-substitute-for-summer.html
790. https://allianceforscience.cornell.edu/blog/2014/10/the-story-of-rainbow-papaya-why-public-sector-biotechnology-research-matters/
791. http://www.innatepotatoes.com/
792. https://www.wired.com/story/genetically-modified-arctic-apple-targets-consumers-not-farmers/
793. https://www.arcticapples.com/how-did-we-make-nonbrowning-apple/
794. https://medium.com/@gmoanswers/nonbrowning-gmo-apples-42d655eccb18
795. https://medium.com/@gmoanswers/nonbrowning-gmo-apples-42d655eccb18
796. https://gmoanswers.com/nonbrowning-gmo-apples-concept-produce-shelves
797. https://www.prnewswire.com/news-releases/intrexon-completes-acquisition-of-okanagan-specialty-fruits-300068285.html
798. https://www.fda.gov/food/newsevents/constituentupdates/ucm533075.htm
799. http://www.odditycentral.com/pics/genetically-modified-pink-pineapples-are-coming-to-a-grocery-store-near-you.html
800. https://www.ncbi.nlm.nih.gov/pmc/articles/PMC1280366/
801. https://www.cbsnews.com/news/what-have-they-done-to-our-food-28-02-2001/

802. https://www.motherearthnews.com/real-food/adding-a-fish-gene-into-tomatoes-zmaz00amzgoe  and https://www.activistfacts.com/person/1487-jane-rissler/

803. https://www.motherearthnews.com/real-food/adding-a-fish-gene-into-tomatoes-zmaz00amzgoe

804. https://www.nature.com/articles/328033a0

805. http://sitn.hms.harvard.edu/flash/2015/insecticidal-plants/

806. https://www.sourcewatch.org/index.php/Bt_Crops

807. http://advances.sciencemag.org/content/2/8/e1600850.full

808. https://www.epa.gov/regulation-biotechnology-under-tsca-and-fifra/framework-delay-corn-rootworm-resistance

809. https://www.ncbi.nlm.nih.gov/pmc/articles/PMC4906537/

810. http://www.plantformcorp.com/science-of-biopharming.aspx

811. https://www.popsci.com/scitech/article/2003-03/something-funny-down-pharm

812. https://www.organicconsumers.org/old_articles/ge/Pharmageddon120502.php

813. http://www.nytimes.com/2002/12/07/us/spread-of-gene-altered-pharmaceutical-corn-spurs-3-million-fine.html and https://geneticliteracyproject.org/2016/01/20/early-struggles-biopharming-poised-make-big-impact-medicine/

814. https://www.elelyso.com/how-is-elelyso-made

815. https://www.elelyso.com/what-is-gaucher-disease

816. http://www.mdpi.com/1422-0067/16/12/26122/htm#table_body_display_ijms-16-26122-t002 and http://time.com/3457472/see-how-ebola-drugs-grow-in-tobacco-leaves/

817. http://www.nejm.org/doi/full/10.1056/NEJMoa1604330

818. https://www.statnews.com/2016/10/12/ebola-zmapp-trial-results/

819. http://www.mdpi.com/1422-0067/16/12/26122/htm#table_body_display_ijms-16-26122-t002

820. http://www.isaaa.org/resources/publications/briefs/49/infographic/default.asp

821. https://www.ers.usda.gov/data-products/adoption-of-genetically-engineered-crops-in-the-us.aspx

822.  https://www.centerforfoodsafety.org/issues/311/ge-foods/about-ge-foods and https://www.news.iastate.edu/media-review/2016/07/08/health%20benefits

823.  http://www.agbioforum.org/v6n12/v6n12a13-carter.htm

824.  https://www.fooddive.com/news/usda-on-gmo-labeling-law-still-on-track-but-a-little-behind/444383/

825.  https://www.npr.org/sections/thesalt/2016/07/14/486060866/congress-just-passed-a-gmo-labeling-bill-nobodys-super-happy-about-it

826.  http://www.scoop.co.nz/stories/PO0205/S00140.htm and http://91.121.143.96/ga2017/assets/resources/WBR%20Rolleston%20-%20CV.PDF

827.  https://www.revolvy.com/main/index.php?s=Lenape&item_type=topic

828.  https://www.revolvy.com/main/index.php?s=Lenape%20(potato)

829.  https://io9.gizmodo.com/the-potato-that-killed-1634775205

830.  https://boingboing.net/2013/03/25/the-case-of-the-poison-potato.html

831.  https://www.revolvy.com/main/index.php?s=Lenape%20(potato)

832.  https://www.nap.edu/read/23395/chapter/2 and https://www.forbes.com/sites/bethhoffman/2013/07/02/gmo-crops-mean-more-herbicide-not-less/#4dfd14323cd5

833.  https://www.roundupreadyxtend.com/About/Traits/Pages/default.aspx

834.  https://www.nytimes.com/2017/09/21/business/monsanto-dicamba-weed-killer.html

835.  https://www.worldwildlife.org/species/monarch-butterfly

836.  http://onlinelibrary.wiley.com/doi/10.1111/j.1752-4598.2011.00142.x/pdf

837.  https://monarchjointventure.org/get-involved/create-habitat-for-monarchs

838.  http://onlinelibrary.wiley.com/doi/10.1111/j.1752-4598.2011.00142.x/pdf and http://www.politifact.com/truth-o-meter/

statements/2015/feb/17/peter-defazio/are-gmos-causing-monarch-butterflies-become-extinc/

839.   https://www.ers.usda.gov/data-products/adoption-of-genetically-engineered-crops-in-the-us/recent-trends-in-ge-adoption.aspx#.UdGtl8u9KSM

840.   https://www.nap.edu/read/23395/chapter/3#14

841.   http://www.pewinternet.org/2016/12/01/public-opinion-about-genetically-modified-foods-and-trust-in-scientists-connected-with-these-foods/

842.   http://beyond-gm.org/wp-content/uploads/2017/04/BGM_StirThePot_Restaurant-survey_FINAL.pdf

843.   http://onlinelibrary.wiley.com/doi/10.1002/j.2326-1951.1976.tb01239.x/abstract;jsessionid=800FD0BE63CAF5F5868E69BA2405A0C7.f04t04

844.   https://www.newyorker.com/news/daily-comment/the-seed-wars

845.   Wilson, Edward O. 2003. How much is the biosphere worth? In The Future Of Life, 114. New York, NY: Vintage Books.

846.   https://genomebiology.biomedcentral.com/articles/10.1186/s13059-015-0607-3 and https://genomebiology.biomedcentral.com/articles/10.1186/s13059-017-1214-2

847.   http://www.ukabc.org/gaiam2_2.htm

848.   http://www.gmwatch.org/en/news/archive/2003/2704-quotes-from-members-of-the-independent-science-panel-on-gm

# THE LAND OF OZ

849.   http://www.wendyswizardofoz.com/script10.htm and https://www.youtube.com/watch?v=lN75xqpqCGE

850.   http://www.thewizardofozblog.com/2012/09/now-thats-a-horse.html

851.   http://www.hammiverse.com/instructionalunits/molecular geneticsadvan/homework/genomewarrior.html and http://libgallery.cshl.edu/items/show/52635

852.    http://environment-ecology.com/gaia/72-james-lovelock-gaia.html

853.    http://www.growtest.org/vandana-shiva-on-food-justice/

854.    For instance, https://www.youtube.com/watch?v=eCPtDVnaQ1w

855.    http://www.who.int/mental_health/en/

856.    http://www.cell.com/cell/abstract/S0092-8674(16)31743-3

857.    https://phys.org/news/2017-01-keys-behavior-tucked-deep-vertebrate.html

858.    http://www.nature.com/news/laser-used-to-control-mouse-s-brain-and-speed-up-milkshake-consumption-1.20995 and https://spectrum.ieee.org/biomedical/devices/neuroscientists-wirelessly-control-the-brain-of-a-scampering-lab-mouse and https://www.technologyreview.com/s/428622/scientists-control-monkeys-brains-with-light/

859.    https://spectrum.ieee.org/biomedical/devices/neuroscientists-wirelessly-control-the-brain-of-a-scampering-lab-mouse

860.    For instance, https://www.slideshare.net/damarisb/transgenic-animals-27039475 and http://www.whatisbiotechnology.org/index.php/science/summary/transgenic

861.    https://blogs.scientificamerican.com/news-blog/fda-approves-blood-thinner-atryn-ma-2009-02-09/

862.    https://www.nature.com/news/us-government-approves-transgenic-chicken-1.18985

863.    https://www.washingtonpost.com/news/speaking-of-science/wp/2017/08/04/gmo-salmon-caught-in-u-s-regulatory-net-but-canadians-have-eaten-5-tons/?utm_term=.ea67b2025412

864.    https://www.eurekalert.org/pub_releases/2014-04/aaft-tae042414.php

865.    https://www.eurekalert.org/pub_releases/2014-04/aaft-tae042414.php

866.    https://www.technologyreview.com/s/545106/human-animal-chimeras-are-gestating-on-us-research-farms/

867.    https://www.technologyreview.com/s/545106/human-animal-chimeras-are-gestating-on-us-research-farms/

868. https://www.technologyreview.com/s/545106/human-animal-chimeras-are-gestating-on-us-research-farms/

869. https://www.technologyreview.com/s/545106/human-animal-chimeras-are-gestating-on-us-research-farms/

870. https://www.gatesnotes.com/Health/Most-Lethal-Animal-Mosquito-Week

871. https://www.gatcsnotes.com/Health/Most-Lethal-Animal-Mosquito-Week

872. https://entomologytoday.org/2014/04/22/x-rays-can-be-used-to-sterilize-mosquitoes/ and https://projects.jsonline.com/news/2017/10/5/mosquito-battle-gets-political.html

873. https://projects.jsonline.com/news/2017/10/5/mosquito-battle-gets-political.html

874. http://www.oxitec.com/our-solution/technology/the-science/

875. http://www.oxitec.com/our-solution/technology/how-it-works/

876. https://www.npr.org/sections/health-shots/2016/11/20/502717253/florida-keys-approves-trial-of-genetically-modified-mosquitoes-to-fight-zika

877. https://www.regulations.gov/document?D=APHIS-2014-0056-0682

878. https://www.wired.com/2016/11/florida-votes-release-millions-zika-fighting-mosquitos/

879. https://www.wired.com/2016/11/florida-votes-release-millions-zika-fighting-mosquitos/

880. https://projects.jsonline.com/news/2017/10/5/mosquito-battle-gets-political.html

881. https://www.sciencedaily.com/terms/dolly_the_sheep.htm

882. https://www.genome.gov/25020028/cloning-fact-sheet/

883. http://learn.genetics.utah.edu/content/cloning/clonezone/ and http://content.time.com/time/covers/0,16641,19970310,00.html

884. https://www.sciencedaily.com/terms/somatic_cell_nuclear_transfer.htm

885. http://www.cell.com/cell/fulltext/S0092-8674(18)30057-6

886.   https://www.statnews.com/2018/01/24/first-cloned-monkeys-dolly-research/

887.   https://www.washingtonpost.com/news/speaking-of-science/wp/2018/01/24/researchers-clone-the-first-primates-from-monkey-tissue-cells/?utm_term=.99251bcef95a

888.   https://viagenpets.com/

889.   https://animalgeneral.net/Home/AboutClinic

890.   http://pittsburgh.cbslocal.com/2017/11/13/texas-company-cloning-pets/

891.   http://pittsburgh.cbslocal.com/2017/11/13/texas-company-cloning-pets/

892.   https://www.amnh.org/exhibitions/dinosaurs-ancient-fossils-new-discoveries/extinction/mass-extinction/

893.   https://www.britannica.com/science/K-T-extinction

894.   https://cosmosmagazine.com/palaeontology/big-five-extinctions

895.   http://www.newsweek.com/us-korea-search-wooly-mammoth-dna-arctic-race-clone-extinct-beast-686311

896.   https://www.nature.com/articles/nature07446

897.   http://www.newsweek.com/resurrecting-mammoths-gets-one-step-closer-221412

898.   http://www.michaelcrichton.com/jurassic-park/

899.   https://www.newscientist.com/article/2121503-can-we-grow-woolly-mammoths-in-the-lab-george-church-hopes-so/

900.   http://bigthink.com/ideafeed/south-korea-working-hard-to-bring-woolly-mammoth-back-to-life

901.   http://www.bbc.com/earth/story/20170221-reviving-woolly-mammoths-will-take-more-than-two-years

902.   http://www.frontlinegenomics.com/news/15649/u-s-korea-race-clone-extinct-beast-woolly-mammoth/

903.   http://www.bbc.com/earth/story/20170221-reviving-woolly-mammoths-will-take-more-than-two-years

904.   http://www.nationalgeographic.com.au/history/why-did-the-woolly-mammoth-die-out.aspx

905.   http://science.sciencemag.org/content/344/6187/1246752

906. http://www.sciencemag.org/news/2017/02/bringing-extinct-species-back-dead-could-hurt-not-help-conservation-efforts

907. http://www.sciencemag.org/news/2017/02/bringing-extinct-species-back-dead-could-hurt-not-help-conservation-efforts

908. https://www.ted.com/talks/craig_venter_unveils_synthetic_life

909. http://engineeringbiologycenter.org/

910. https://www.theguardian.com/science/2007/oct/06/genetics.climatechange

911. https://medium.com/neodotlifc/6-things-to-watch-in-synthetic-biology-f76666c7114e

912. Chargaff, Erwin. 1978. Heraclitean Fire: Sketches from a Life Before Nature, 183. New York, NY: Rockefeller University Press and https://www.amazon.com/Heraclitean-Fire-Sketches-Before-Nature/dp/0874700299

913. http://science.sciencemag.org/content/192/4243/938.long

## EYE OF THE BEHOLDER

914. http://leethomson.myzen.co.uk/The_Twilight_Zone/The_Twilight_Zone_2x06_-_The_Eye_of_the_Beholder.pdf

915. http://www.newsweek.com/2017/07/07/natural-selection-new-forms-life-scientists-altering-dna-629771.html

916. http://www.ndss.org/about-down-syndrome/down-syndrome/

917. https://www.cbsnews.com/news/down-syndrome-iceland/

918. https://www.ncbi.nlm.nih.gov/pmc/articles/PMC3740159/

919. https://www.aol.com/article/2016/08/17/5-amazing-people-with-down-syndrome-who-do-extraordinary-things/21453609/ and https://www.liveaction.org/news/9-successful-people-with-down-syndrome-who-prove-life-is-worth-living/

920. https://www.youtube.com/watch?v=HwxjoBQdn0s

921. Plato. The Republic, 407 and http://ancientimes.blogspot.com/2017/02/ancient-eugenics-much-more-than-just.html

922. Darwin, Charles. 1866. On the Origin of Species by Means of Natural Selection, 577. London, England: John Murray.

923.  Galton, Francis. 1869. *Hereditary Genius*. Boston, MA: D. Appleton & Company  and https://www.nature.com/scitable/forums/genetics-generation/america-s-hidden-history-the-eugenics-movement-123919444

924.  https://www.etymonline.com/word/eugenics

925.  http://knowgenetics.org/history-of-eugenics/

926.  Charles Darwin to Francis Galton. January 4, 1873. Correspondence.

927.  Wells, H. G. 1904. Eugenics: Its definition, scope, and aims In *The American Journal of Sociology*, Vol. X, No. 1.

928.  https://harvardmagazine.com/2016/03/harvards-eugenics-era

929.  https://www.newyorker.com/books/page-turner/the-forgotten-lessons-of-the-american-eugenics-movement

930.  https://sangerpapers.wordpress.com/2014/02/12/jan-2-1923-first-legal-birth-control-clinic-opens-in-u-s/

931.  https://www.nyu.edu/projects/sanger/webedition/app/documents/show.php?sangerDoc=101807.xml

932.  https://www.biography.com/people/margaret-sanger-9471186

933.  Sanger, Margaret. 1922. *The Pivot of Civilization* New York, NY: Bretano's Publishers.

934.  https://www.nyu.edu/projects/sanger/webedition/app/documents/show.php?sangerDoc=238946.xml

935.  Sanger, Margaret. 1920. The Goal. In *Woman and the New Race*. New York, NY: Bretano's Publishers and http://www.bartleby.com/1013/18.html

936.  Sanger, Margaret. 1922. *Woman, Morality, and Birth Control*, 12. New York, NY: New York Publishing Company.

937.  https://www.ushmm.org/wlc/en/article.php?ModuleId=10008193

938.  Weikart, Richard. 2004. *From Darwin to Hitler: Evolutionary Ethics, Eugenics, and Racism in Germany*. Basingstroke, United Kingdom: Palgrave Macmillan. See https://www.goodreads.com/book/show/54328.From_Darwin_to_Hitler

939.  Turner, Henry Ashby, Jr. 1985. *Hitler—Memoirs of a Confidant*, 145–146. New Haven, CT: Yale University Press and https://

yalebooks.yale.edu/book/9780300032949/hitler-memoirs-confidant

940. https://www.theguardian.com/commentisfree/2016/feb/02/genetic-editing-playing-god-children-british-scientists-embryos-dna-diseases

941. https://www.nature.com/articles/nature23305. NB: As of this writing, there's some argument about how the snipped gene was actually repaired—or whether it was even repaired at all, as claimed by the scientists. See, for example, http://www.sciencemag.org/news/2017/08/skepticism-surfaces-over-crispr-human-embryo-editing-claims

942. https://www.newscientist.com/article/2123973-first-results-of-crispr-gene-editing-of-normal-embryos-released/

943. https://www.npr.org/sections/health-shots/2016/09/22/494591738/breaking-taboo-swedish-scientist-seeks-to-edit-dna-of-healthy-human-embryos

944. https://www.yahoo.com/news/scientists-edit-embryos-genes-study-early-human-development-170515808.html

945. https://news.nationalgeographic.com/2017/08/human-embryos-gene-editing-crispr-us-health-science/ and http://www.ohsu.edu/xd/research/centers-institutes/embryonic-cell-gene-therapy-center/people.cfm

946. https://www.npr.org/sections/health-shots/2016/09/22/494591738/breaking-taboo-swedish-scientist-seeks-to-edit-dna-of-healthy-human-embryos

947. http://content.time.com/time/magazine/article/0,9171,95244,00.html

948. https://www.npr.org/sections/health-shots/2016/09/22/494591738/breaking-taboo-swedish-scientist-seeks-to-edit-dna-of-healthy-human-embryos

949. https://www.wsj.com/articles/in-gene-editing-advance-scientists-correct-defect-in-human-embryos-1501693349?mg=prod/accounts-wsj

950. http://www.sfgate.com/business/article/Scientists-publish-or-perish-credo-now-patent-2891077.php

951. https://www.technologyreview.com/s/603633/us-panel-endorses-designer-babies-to-avoid-serious-disease/

952. https://www.technologyreview.com/s/603633/us-panel-endorses-designer-babies-to-avoid-serious-disease/

953. Rescher, Nicholas. 1984. *The Limits of Science*, 102–103. Pittsburgh, PA: University of Pittsburgh Press and https://www.goodreads.com/book/show/5120050-the-limits-of-science

954. https://news.nationalgeographic.com/2017/08/human-embryos-gene-editing-crispr-us-health-science/

955. https://news.nationalgeographic.com/news/2007/11/071114-AP-monkey-clones.html

956. https://www.npr.org/sections/health-shots/2013/05/15/183916891/scientists-clone-human-embryos-to-make-stem-cells

957. https://news.nationalgeographic.com/2017/08/human-embryos-gene-editing-crispr-us-health-science/

958. https://news.nationalgeographic.com/2017/08/human-embryos-gene-editing-crispr-us-health-science/

959. https://news.nationalgeographic.com/2017/08/human-embryos-gene-editing-crispr-us-health-science/

960. https://www.newscientist.com/article/dn4266-controversial-three-parent-pregnancy-revealed/

961. https://www2.le.ac.uk/departments/emfpu/to-be-deleted/explained/mitochondrial

962. https://publications.nigms.nih.gov/insidelifescience/genetics-numbers.html and http://www.majordifferences.com/2015/05/difference-between-mitochondrial-dna.html#.WkeVylWnGM9

963. https://www2.le.ac.uk/departments/emfpu/to-be-deleted/explained/mitochondrial

964. https://www2.le.ac.uk/departments/emfpu/to-be-deleted/explained/mitochondrial

965. https://rarediseases.info.nih.gov/diseases/6877/leigh-syndrome

966.  http://www.fertstert.org/article/S0015-0282%2816%2962670-5/
      fulltext?rss=yes

967.  https://www.newscientist.com/article/2107219-exclusive-worlds-
      first-baby-born-with-new-3-parent-technique/

968.  https://www.newscientist.com/article/2107219-exclusive-worlds-
      first-baby-born-with-new-3-parent-technique/

969.  https://www.technologyreview.com/s/608033/the-fertility-doctor-
      trying-to-commercialize-three-parent-babies/

970.  https://www.sartcorsonline.com/rptCSR_PublicMultYear.aspx

971.  https://www.technologyreview.com/s/608033/the-fertility-doctor-
      trying-to-commercialize-three-parent-babies/

972.  http://www.rbmojournal.com/article/S1472-6483(16)30439-4/
      fulltext

973.  https://www.technologyreview.com/s/608033/the-fertility-doctor-
      trying-to-commercialize-three-parent-babies/

974.  http://www.fertstert.org/article/S0015-0282%2816%2962670-5/
      fulltext?rss=yes

975.  https://www.nature.com/news/three-parent-baby-claim-raises-
      hopes-and-ethical-concerns-1.20698

976.  https://www.fda.gov/downloads/BiologicsBloodVaccines/
      GuidanceComplianceRegulatoryInformation/Compliance
      Activities/Enforcement/UntitledLetters/UCM570225.pdf?source
      =govdelivery&utm_medium=email&utm_source=govdelivery

977.  https://www.darwinlife.com/

978.  https://www.sciencenewsforstudents.org/article/how-make-three-
      parent-baby and https://www.newscientist.com/article/2108549-
      exclusive-3-parent-baby-method-already-used-for-infertility/

979.  https://www.hfea.gov.uk/about-us/news-and-press-releases/2017-
      news-and-press-releases/hfea-statement-on-mitochondrial-
      donation/ and http://www.newcastle-hospitals.org.uk/hospitals/
      fertilitygenetics_newcastle-fertility-centre.aspx

980.  https://www.technologyreview.com/s/608033/the-fertility-doctor-
      trying-to-commercialize-three-parent-babies/

981.   https://www.theverge.com/2014/3/27/5553044/first-functional-eukaryotic-chromosome

982.   https://medium.com/neodotlife/6-things-to-watch-in-synthetic-biology-f76666c7114e

983.   http://icelandreview.com/news/2012/08/13/iceland-ranks-high-world-atheist-list

984.   http://icelandmag.visir.is/article/00-icelanders-25-years-or-younger-believe-god-created-world-new-poll-reveals

985.   https://www.cbsnews.com/news/down-syndrome-iceland/

986.   Sandel, Michael J. 2004. The case against perfection. *The Atlantic Monthly* April. See https://www.theatlantic.com/magazine/archive/2004/04/the-case-against-perfection/302927/

987.   http://breakpoint.org/2004/04/building-better-babies/

988.   https://www.cbsnews.com/news/down-syndrome-iceland/

989.   Harari, Yuval Noah. 2014. *Sapiens: A Brief History of Humankind*. Oxford, England: Signal. See https://www.amazon.com/Sapiens-A-Brief-History-Humankind/dp/077103850X

## CONCLUSION

## SCIENCE, SACREDNESS, AND SELF-DESTRUCTION

990.   Schaar, John H. 1981. *Legitimacy In The Modern State*, 321. Piscataway, NJ: Transaction Publishers and https://www.goodreads.com/quotes/279924-the-future-is-not-some-place-we-are-going-but

991.   https://www.amazon.com/Encyclopedia-Wars-Facts-Library-History/dp/0816028516. Also, see https://www.huffingtonpost.com/rabbi-alan-lurie/is-religion-the-cause-of-_b_1400766.html and https://carm.org/religion-cause-war

992.   Dawkins, Richard. 2008. *The God Delusion. Boston*, MA: Mariner Books. See https://www.amazon.com/God-Delusion-Richard-Dawkins/dp/0618918248

993. http://www.age-of-the-sage.org/quotations/quotes/richard_dawkins_god.html

994. https://www.hinduismtoday.com/modules/smartsection/item.php?itemid=5561

995. https://www.hinduismtoday.com/modules/smartsection/item.php?itemid=5561

996. Einstein, Albert. 1987. *Letters To Solovine:* 1906–1955, 131. New York, NY: Philosophical Library. See: https://www.amazon.com/Letters-Solovine-1906-1955-Albert-Einstein/dp/1453204881

997. http://www.pewforum.org/2017/04/05/the-changing-global-religious-landscape/

998. http://www.pewresearch.org/fact-tank/2016/06/01/10-facts-about-atheists/

999. http://www.dailymail.co.uk/news/article-4973674/Collective-consciousness-replace-God-author-Dan-Brown.html

1000. http://www.pewresearch.org/fact-tank/2017/04/07/why-people-with-no-religion-are-projected-to-decline-as-a-share-of-the-worlds-population/

1001. http://religion.blogs.cnn.com/2011/05/12/religious-belief-is-human-nature-huge-new-study-claims/

1002. https://phys.org/news/2011-05-humans-predisposed-gods-afterlife.html

1003. https://phys.org/news/2011-05-humans-predisposed-gods-afterlife.html

1004. Einstein, Albert. 2007. *The World As I See It*, 91. San Diego, CA: Book Tree and https://archive.org/stream/AlbertEinstein TheWorldAsISeeIt/The_World_as_I_See_it-AlbertEinstein UpByTj_djvu.txt

1005. http://en.radiovaticana.va/storico/2013/07/17/pope_francis_all_life_has_inestimable_value/en1-711052

1006. http://www.ucmp.berkeley.edu/history/aristotle.html

1007. http://www.ibnalhaytham.com/

1008. any references available. For a readable one, see https://science .howstuffworks.com/innovation/scientific-experiments/scientific-method3.htm

1009. http://blogs.nature.com/soapboxscience/2011/05/18/science-owes-much-to-both-christianity-and-the-middle-ages

1010. Many references available. See, for example, http://blogs.nature .com/soapboxscience/2011/05/18/science-owes-much-to-both-christianity-and-the-middle-ages

1011. https://blogs.scientificamerican.com/cross-check/is-scientific-materialism-almost-certainly-false/

1012. In 1995 on the TV show *Reality on the Rocks: Beyond Our Ken*, Hawking claimed "the human race is just a chemical scum on a moderate size planet, orbiting round a very average star in the outer suburb of one among a billion galaxies."

1013. https://www.aaas.org/page/what-scientism

1014. https://twitter.com/ericidle/status/842157775115452416?lang=en

1015. http://www.patheos.com/blogs/friendlyatheist/2016/04/15/does-bill-nye-really-want-to-put-climate-change-deniers-in-jail/

1016. For instance, see http://econfaculty.gmu.edu/wew/articles/07/ silencingdissent.htm and https://www.epw.senate.gov/public/ index.cfm/press-releases-all?ID=ac856964-802a-23ad-45e0-3e1e9935490b

1017. http://www.foxnews.com/opinion/2017/04/27/has-science-lost-its-way.html. See also, for instance, http://reason.com/ archives/2016/08/26/most-scientific-results-are-wrong-or-use

1018. https://plato.stanford.edu/entries/aquinas/#ThoAri

1019. For a comprehensive history, see, for example Gingerich, Owen. 2005. *The Book Nobody Read: Chasing the Revolutions of Nicolaus Copernicus*. London, United Kingdom: Penguin and https://www.amazon.com/Book-Nobody-Read-Revolutions-Copernicus/dp/B000BNPG8C. For a brief, simple-to-read treatment of the subject, see http://physics.ucr.edu/~wudka/ Physics7/Notes_www/node41.html or https://www.khanacademy

.org/partner-content/big-history-project/big-bang/how-did-big-bang-change/a/nicolaus-copernicus-bh

1020. https://www.catholic.com/tract/the-galileo-controversy

1021. For more about my religious awakening, see *Can a Smart Person Believe in God?* and *Amazing Truths: How Science and the Bible Agree.* https://www.amazon.com/Can-Smart-Person-Believe-God/dp/0785287892 and https://www.amazon.com/Amazing-Truths-Science-Bible-Agree/dp/0310343755/ref=sr_1_1?s=books&ie=UTF8&qid=1515168413&sr=1-1&keywords=amazing+truths+how+science+and+the+bible+agree

1022. http://www.foxnews.com/opinion/2017/05/08/why-stephen-hawkings-dire-warning-is-all-wrong-about-humans-and-earth.html

1023. https://opinionator.blogs.nytimes.com/2012/06/10/are-we-living-in-sensory-overload-or-sensory-poverty/

1024. Quote is from Mark 8:36. For information about John Mark, see https://www.gotquestions.org/John-Mark-in-the-Bible.html

# NAME INDEX

## A

Acemoglu, Daron, 100
Ackerman, Diane, 264
Adams, Scott, 129
Aiken, Clay, 33
Akbari, Anna, 40
Al-Assad, Bashar, 71
Alhazen, 259
Al Mahdi, Rabab, 60
Al Razooqi, Brigadier Khalid Nasser, 103
Altman, Sam, 137, 147
Anderson, Mac, 34
Angle, Colin, 168–169
Annas, George, 242–243
Ann-Margaret, 33
Aquinas, Thomas, 262
Aristotle, 258
Armstrong, Neil, 38, 132
Arness, James, 203
Asimov, Isaac, 134–135
Assad, Asma, 71
Assael, Yannis, 109
Assange, Julian, 177
Atanasoff, John Vincent, 24
Atick, Joseph J., 187
Austin, Greg, 147
Avery, Oswald, 200
Axelrod, Alan, 254–255

## B

Babbage, Charles, 24
Bacon, Francis, 259
Bacon, Roger, 259
Banzhaf, John, 121
Barrat, James Rodman, 135
Barros, Eduardo, 184
Bartlett, Jamie, 135–136
Beane, Matt, 109–110
Beckett, Miles, 67, 70
Bedoya, Alvaro, 156
Bell, Alexander Graham, 22
Bennett, Joseph, 231
Berg, Paul, 242
Berners-Lee, Tim, 27–28, 32, 65, 76
Beroukhim, Raffie, 150
Bezos, Jeff, 15, 35–36, 118
Bieber, Justin, 34
Bing, Xu, 152
Blanchett, Cate, 116
Boeke, Jef, 247
Boone, Pat, 33
Boren, Braxton, 22
Boulmalf, Mohammed, 166–167
Bowes, Edward "Major," 32
Brandeis, Louis, 145–146
Branson, Richard, 15
Briggs, Patricia, 31
Brokaw, Tom, 204

335

# SUBJECT INDEX